# Fundamentals of Mobile Computing

# Fundamentals of Mobile Computing

Nicolas Hudson

MURPHY & MOORE
www.murphy-moorepublishing.com

Fundamentals of Mobile Computing
Nicolas Hudson
ISBN: 978-1-63987-249-7 (Hardback)



MURPHY & MOORE
Published by Murphy & Moore Publishing,
1 Rockefeller Plaza,
New York City, NY 10020, USA

**Cataloging-in-Publication Data**

Fundamentals of mobile computing / Nicolas Hudson.
p. cm.
Includes bibliographical references and index.
ISBN 978-1-63987-249-7
1. Mobile computing. 2. Context-aware computing. 3. Portable computers.
4. Electronic data processing. I. Hudson, Nicolas.
QA76.59 .F86 2022
004--dc23

# Table of Contents

# Preface

Human-computer interactions in which computers are expected to be transported during the course of usage are defined as mobile computing. It allows wireless transmission of data, voice and video without the need of a fixed physical link. GSM, CDMA and LTE are some of the wireless technologies used by these devices. The guiding principles of mobile computing technologies are portability, connectivity, interactivity and individuality. Some of the examples of common mobile computing devices are portable computers, wearable computers, smart cards, personal digital assistants and laptops. This domain has greatly benefitted from the technical advancements in computing technology such as low-power PC processors, small size digital memory technology, and inexpensive display systems. Some of the security issues faced by mobile computing technologies include hacking, piracy, online frauds and industrial espionage. Mobile computing is an upcoming field of science that has undergone rapid development over the past few decades. Some of the diverse topics covered herein address the varied branches that fall under this category. This book is an essential guide for both academicians and those who wish to pursue this discipline further.

A detailed account of the significant topics covered in this book is provided below:

Chapter 1- Mobile computing is a technology that allows the transmission of data, voice and video through human-computer interaction without the need of a physical link. Computing devices such as smart cards, mobile phones, portable computers, portable digital assistants, etc. are used to conduct mobile computing. This chapter sheds light on mobile computing and its various devices for a thorough understanding of the subject.

Chapter 2- Mobile communication refers to the technology that allows us to communicate with others in an easier way without the need of a physical connection. High capacity load balancing, scalability, network management, access controls, roaming, modulation, antennas and propagation are some features of mobile communications which are discussed comprehensively in this chapter.

Chapter 3- A medium access control is a network data transfer policy responsible for the flow control and data transmission through a wired, optical or wireless transmission medium. It ensures non-collision and eases transfer of data packets between two terminals. FDMA, TMA, CDMA and SDMA are some of the medium access schemes used in medium access control. This chapter touches on the subject of medium access control and medium access schemes in detail.

Chapter 4- A mobile network is a radio network with wireless last links distributed over land areas and served by cell sites or base stations. The transport layer is a conceptual division of methods in the layered architecture of protocols in the network stack in the OSI model and the internet protocol suite. Mobile internet protocol and transport control protocol are the protocols used in mobile networks. This chapter has been carefully written to provide an easy understanding of mobile networks and transport layers.

Chapter 5- A cellular network is a mobile network that provides services over land through cells and base stations. Cellular networks provide a larger coverage area with increased capacity over alternative solutions. Cellular network technology has come a long way from 2G to 3G and 4G with 5G being the latest introduction. This chapter closely examines the various cellular networks to provide an extensive understanding of the subject.

Chapter 6- Mobile security refers to the measures designed to protect the information stored on and transmitted by computing devices from unauthorized access and hacking. With the usage of computing devices concerns have risen over the security afforded by computing devices. This chapter focuses on mobile security and its related concerns at length for the benefit of the reader.

It gives me an immense pleasure to thank our entire team for their efforts. Finally in the end, I would like to thank my family and colleagues who have been a great source of inspiration and support.

**Nicolas Hudson**

1

# Mobile Computing and Computing Devices

Mobile computing is a technology that allows the transmission of data, voice and video through human-computer interaction without the need of a physical link. Computing devices such as smart cards, mobile phones, portable computers, portable digital assistants, etc. are used to conduct mobile computing. This chapter sheds light on mobile computing and its various devices for a thorough understanding of the subject.

## Mobile Computing

Mobile Computing is a technology that provides an environment that enables users to transmit data from one device to another device without the use of any physical link or cables.

In other words, you can say that mobile computing allows transmission of data, voice and video via a computer or any other wireless-enabled device without being connected to a fixed physical link. In this technology, data transmission is done wirelessly with the help of wireless devices such as mobiles, laptops etc.

This is only because of Mobile Computing technology that you can access and transmit data from any remote locations without being present there physically. Mobile computing technology provides a vast coverage diameter for communication. It is one of the fastest and most reliable sectors of the computing technology field.

The concept of Mobile Computing can be divided into three parts:

- Mobile Communication.
- Mobile Hardware.
- Mobile Software.

### Mobile Communication

Mobile Communication specifies a framework that is responsible for the working of mobile computing technology. In this case, mobile communication refers to an infrastructure that ensures seamless and reliable communication among wireless devices. This framework ensures the consistency and reliability of communication between wireless devices. The mobile communication framework consists of communication devices such as protocols, services, bandwidth, and portals necessary to facilitate and support the stated services. These devices are responsible for delivering a smooth communication process.

Mobile communication can be divided into following four types:

- Fixed and Wired.
- Fixed and Wireless.
- Mobile and Wired.
- Mobile and Wireless.

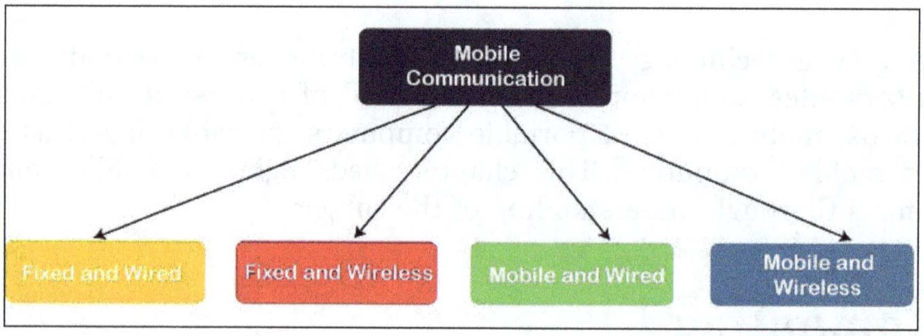


Fixed and Wired: In Fixed and Wired configuration, the devices are fixed at a position, and they are connected through a physical link to communicate with other devices. For example, Desktop Computer.

Fixed and Wireless: In Fixed and Wireless configuration, the devices are fixed at a position, and they are connected through a wireless link to make communication with other devices. For example, Communication Towers, WiFi router.

Mobile and Wired: In Mobile and Wired configuration, some devices are wired, and some are mobile. They altogether make communication with other devices. For example, Laptops.

Mobile and Wireless: In Mobile and Wireless configuration, the devices can communicate with each other irrespective of their position. They can also connect to any network without the use of any wired device. For example, WiFi Dongle.

## Mobile Hardware

Mobile hardware consists of mobile devices or device components that can be used to receive or access the service of mobility. Examples of mobile hardware can be smartphones, laptops, portable PCs, tablet PCs, Personal Digital Assistants, etc.

These devices are inbuilt with a receptor medium that can send and receive signals. These devices are capable of operating in full-duplex. It means they can send and receive signals at the same time. They don't have to wait until one device has finished communicating for the other device to initiate communications.

## Mobile Software

Mobile software is a program that runs on mobile hardware. It is designed to deal capably with the characteristics and requirements of mobile applications. It is the operating system for the appliance of mobile devices. In other words, you can say it the heart of the mobile systems. This is an essential component that operates the mobile device.

It provides portability to mobile devices, which ensures wireless communication. In 1894, Marconi done experiment along with Hertzian rays and identify the waves over very long distance. Then developed the first wireless telegraph company.

- 1980 – Developed first consumer Laptop.
- 1981 – Developed portable Computer by Epson.
- 1984 – First touch screen laptop was developed.
- 1990 – New portable laptop released by Apple with 640*640 display screen.
- 1992 – Released standard Windows 3.1.1 Operating System.
- 1993 – Developed first PDA by Apple.
- 2007 – Launched first I Phone along with touch screen.
- 2009- Launched first smart phone with Android.
- 2011 – Start evolution of Mobile Computing.

## Future of Mobile Computing

The enhancement in Artificial Intelligence, increasing the speed to computing terminals So the future of Mobile computing is more brightness.

Increasing the technology in portability and light weighted devices, so as per this statement mobile computing devices are improving in portability and small in size.

Now these days, most of the people have increasing laziness, they want to perform all the task while sitting in own house. so mobile computing fulfill this are because you can done mostly all task on mobile computing device like as want to shop then no require physically visit at any store, you can visit any e-commerce website and make your order.

As per research of this year that 80% the mobile devices converted into smart phone, because mobile company are working as per future trend to mobile computing.

Now in conclusion, Mobile Computing future is most brightness because all user and companies are adopting mobile computing.

## Features of Mobile Computing

- Easy to handle and carry these small devices.
- Data can be transferred easily between users.
- Collect simulated data to current zone or your time.
- Arbitrary network, easily connect to other environment and transmit data.
- Having fast processor speed.
- Good battery life.
- Huge memory capacity.
- Bar code scanner.
- Windows operating system.
- Camera.
- Global positioning system.
- Durability.

## Mobile Computing Applications

There are different types of real life applications which are used in the Mobile computing, such as:

### Traffic

During traveling in traffic if we require to know road situation, latest news and when if feel more stress in driving then can play music and other important broadcast data are received through digital audio broadcasting(DAB). If we forget the road then we can know our exact location with the help of global positioning system (GPS).In case if got accident then can to inform police and ambulance via an emergency call to the service provider, which help to improve organization and save time & money.

### Emergencies Situation

To play a vital role in the medical sector can hire an ambulance with great quality wireless connection and help of this can carry significant information about injured persons. The useful step can prepare for a particular accident and doctor can be consulted for diagnosis. Only Wireless networks work of communication in nature disaster 2 such as earthquakes, tsunami, flood, and fire. In worst conditions only decentralized, wireless ad-hoc networks survive. Means that can handle Emergencies situation by mobile computing easily.

## Use in Business

As per business point of view CEO help of this computing system can represent the presentation at the front of their clients while can access hot news of the market. Help of video conference could be discuss at the topic without hindrance any time. Another side if traveling salesman wants to access the company database as per requirement then can be retrieved data on his wireless device and maintain the consistency company's database. Cause of these every employee are updated up to date.

## Credit Card Verification

Credit card verification using this computing most secure. In respect of Sale terminals(POS) when customer buy items in malls and other small shops when and pay bill in the form of swap credit card for transactions then need to establish network in between POS terminal and bank central computer then over protected cellular network verify the credential information of card fastly, if match it then proceed further otherwise denied get boost up speed of transaction process and relieve the burden at the POS network.

## Replacement of Fixed Networks

Wired network has been replaced in wireless network e.g. trade shows, remote sensors and historical buildings. in wired networks, weather forecasting, earthquake detection and to get environmental data are impossible. This is possible only in adapting the replacement of fixed networks in this computing.

## Infotainment

Wireless networks are capable to deliver the latest information at any suitable regions and can download knowledge about concert at morning through wireless network that concert is conducting in any region as well as another growing field of wireless network applications lies in entertainment and games to enable, e.g., ad-hoc gaming networks as soon as people meet to play together. So infotainment by wireless computing is more easy.

## Mobile Computing Services

Software as a service is best platform with the help of rendering the many utilities and applications on the internet world in the form of services. Cloud Computing help to optimize your resources and scale up while control your cost. Suggest the best mobile hardware and software for your using utilities and applications.

In arbitrary wireless network service help to connect freely network, and with the help of this concept, you can access data anywhere without time and place limitation.

Mobile terminal phases are:

- Provide the current and appropriate O/S and other software.

- Examine the loading time of particular software.
- Setup the wireless networks such as WAN and LAN.

## Types of Mobile Computing Devices

There are many types of Mobile Computing devices that permit to user for accessing all type of data and information from anytime and anywhere without boundation of time and place. Today, these devices are more demanded in market because having small size and light weight.

There are two types of mobile computing, such as:

- Portable Computing.
- Mobility Computing.

## Mobile Computing Architecture and Structure

Mobile Computing has 3-tier architecture with various its functions, in which include major three segments are:

- Data access layer.
- Application tire in middle line.
- Client tier on the surface.

Each tier is spitted in many network and not require any physical location on all terminals on the network.

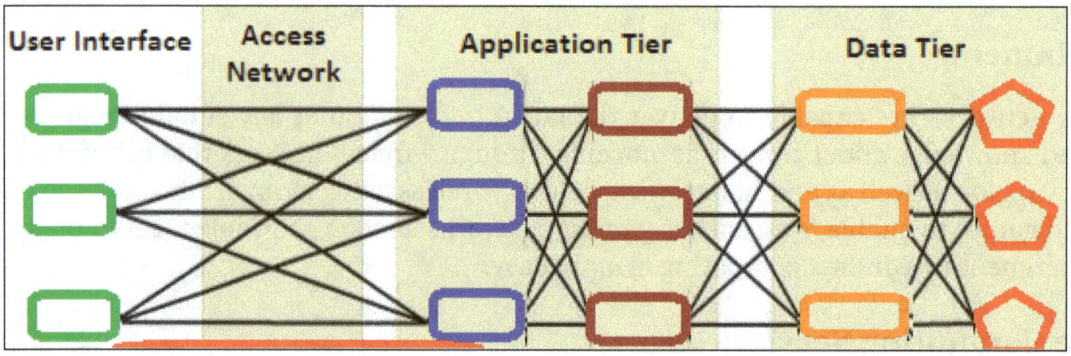


## Presentation Layer

This layer provides data to the user and allows to data feeding and manipulation, and proceeds all data as business layer.

## Business Logic Layer

Business logic work as a server for user requests from work point. This works according to bringing or entering business rules through the data layer.

## Data Access Layer

This layer work as DBMS access layer because this layer provides data to other two layers. Allow dependencies on the storage area can be done updation.

## Mobile Computing Security

Two types attacks are existed in the Mobile Computing security:

- Passive attack – In this attack intruder only monitor the transmitted data over the network.
- Active attack – In which intruder gives some modification in the original data.

Concern Security is:

- Encrypted data.
- Outsource detection.
- Lack of data capacity.
- Terminal tracer.
- Set up timeout system.
- Use reliable sources such as Google Play and Amazon App.

## Disadvantages of Mobile Computing

There are many limitations and challenges in Mobile Computing. Explain below:

- Battery consumption hindrance.
- Interference is persisted in shielding.
- Inefficient bandwidth in transmission.
- Connection losses over entithe re network.
- Network stability.
- Interoperability problem.
- Protection constraints.

## Advantages of Mobile Computing

Now these days, Mobile Computing has more need because there are various benefits such as:

- Increasing Productivity – Mobile devices can be used in the field of various instruments, so reducing time and cost for customers and themselves.
- Entertainment – For entertainment purposes, mobile computing devices can be used for both people and customers for personal and even presentations.

- Portability – The main advantages of this, you are not bound to one Zone. It helps to access any wireless devices without place boundation.
- Cloud Computing – This service is available to save documents on online servers and whenever you make a connection to the Internet, they are able to access them at any time and anywhere and these files can be used for many mobile devices or even that can also access on PC.

## Mobile Computing Devices

### Smart Cards

We live in a smart world where technological trends and advancements are making life easier for everyone. Making payments and shopping online across various platforms has never been this swift and secured in the 21st century. Smart card technology is the fuel driving this convenient means of making payments and carrying out online transactions.

Smart cards are now common in every economy across the world - they are being used by almost everyone that receives a paycheck. The idea of a smart card is to reduce the archaic manner in which people carry cash about and feel insecure owing to prying eyes and numerous cases of heists associated with moving large chunks of cash about. Essentially, smart cards are providing a safer and secure means of conducting financial transactions while making life easier in a world that is going cashless.

### What Smart Card Technology is all About?

Smart card technology involves making a tiny rectangular piece of card smart. These cards are tiny enough to fit in your wallet or back pockets, and that's where the first convenience it affords can be seen.

Various institutions ranging from banks, companies, business enterprises, and many other institutions make use of smart cards for different purposes. The cards can come in different sizes, but they all share one thing in common; the technology that drives them.

This smart technology enables users to store specific information unique to them which they can use to make payments and other transactions without involving physical cash.

## Why are they Labeled 'Smart Cards'?

They are called smart cards because they aren't just cards; they are cards bearing a substantial amount of private information and data limited to a particular user: the owner of the card.

These cards have memory chips embedded in them that are microprocessors with the capacity to store a lot of information about its users. These users can utilize the technology to carry out online financial transactions such as shopping without cash in hand, making payments like electric bills, and transferring funds to a second party.

With microprocessors, the cards are now smart because they are capable of housing information and making use of them as the occasion demands.

## How Smart Cards are Constructed

The construction of a smart card entails four basic steps: designing, fabrication, coding, and data loading.

## Designing

This is the first step which requires the designer or programmer to allocate a memory size to the chip, specify the clock speed, the volatile memory types, and the operating system.

It also requires the programmer to create application software for the card while specifying the card type and any other features he might want to include in the card.

## Fabrication of the Chip

The chip made of silicon is being fixed into the card at this second stage. This silicon chip is being linked to connectors with connecting wires either by soldering them or by bounding them together.

After this, the chip on the board substrate is sealed with an epoxy resin and glued directly to the substrate of the card. The substrate here is the plastic which can be made of polyvinyl chloride (PVC) or any other synthetic plastic.

## Coding

This is an integral step where the foundation for the functionality of the card is laid. In this step, the codes are being entered into the chip memory using special commands.

### Data Loading

This is the stage where the intended user's personal data is being loaded into the memory chip.

### How the Smart Cards Work

Smart cards are like the regular cards you can find around you (the same size and made of the same plastic material) but they have a memory chip contained in a contact pad.

The contact pad bearing the embedded memory chip is the basis for which a transaction can be made using a smart card. Once this contact pad is detached, the card is no longer smart.

Smart cards don't work alone - they need a smart card reader to function. When certain relevant personal information has been written into the smart card by its issuers, you'll need a card reader where you can slot in the card, punch in your personal security code, and carry out any transaction.

It is the contact pad that comes in real contact with the card reader and subsequently establishes an electronic interaction between itself and the card reader.

This then enables you to make payments via a Point of Sale (POS) system or other mediums to carry out transactions. However, some smart cards don't require a card reader interface to function - contactless cards are increasing in popularity due to their convenience.

### Categories of Smart Cards

Smart cards are classified based on the mechanism of the cards and their configuration.

### Classification based on Mechanism

This classification is based on the workings of the card and how they are being used. Under this category we have:

### Contact Smart Card

This is about the most common smart card in existence. Most credit cards, ATM cards, and SIM cards fall into this category.

Smart contact cards work by inserting it into a card reader and with the contact pad bearing a memory chip where information is being stored, the card reader can read whatever is written on the card and allow the user perform any transaction he intends doing.

They are called smart cards because they come in contact with a medium before they can function. The medium, in this case, is any digital interface like the smart card reader that allows for the transmission of information from the card to the reader.

### Contactless Smart Card

Just like the name implies, it's a contactless smart card and does not need to come in any physical contact with an interface before you can make use of it.

There are no card readers or digital interfaces to create a connection and subsequent transaction in the case of contactless smart cards. These kinds of cards make use of radio frequencies or Near Field Communication (NFC) technology that establishes a wireless connection between the card and wherever you intend using it.

The proximity contactless smart card is a typical example. The proximity card has only a sizeable memory space can only be read and not written to. It means you can't save anything on this card; you can only use the configuration on it to access restricted places, devices or gain access into a building.

If you are trying to access a building, all you do is to flash the card at the interface in the automated security door and you are granted access after your personal data have been accessed and verified.

### Hybrid Cards

The hybrid cards are a powerful aggregation of both types of cards in this category. These cards can function as a contact smart card and also as a contactless smart card as the occasion demands.

They have both technologies embedded in them to function in such dual capacities. These cards quite rare, existing in a class of their own.

## Classification based on Configuration

### Memory Cards

These are smart cards that are only for a temporal basis. Their memory capacities are extremely low which is why their functions are quite limited to certain activities or usage like reading, writing data and storing a limited amount of information required for a peculiar function.

Whatever data that is written on them can't be edited or changed. It is static and can only perform one job description at a time after which it loses its relevance and most times it's been discarded after usage.

### Microprocessor Cards

The microprocessors are like mini-computers that are being made so portable that they can be hidden in your pockets. These smart cards have larger memory spaces to accommodate tons of data.

The processors are just like your traditional computers and are made to work using software (operating systems), so you can read information on them, write to them, rewrite a written data, edit and manipulate whatever they contain at will.

## Mobile Phones

A mobile phone is a portable electronic device for communication purposes; it offers the user to carry mobile anywhere the mobile has network coverage. The development of mobile phone is huge in technology and usage. Mobile phone have made the whole communication comes into our hand and helps user for easy use.

As of much of last century radio telephony was used what we now refer as mobile phone or cell phone. The first mobile phone was invented in 1973, even though the concept of the cellular networks where understood in 1940s.

Telephone a dream invention of human has been easily taken off by the mobile phone. Mobile gives us free to carry anywhere but the landline telephones lacks. Mobile is a modern communication device which can be used for calling a person on other end, SMS, Video calling, browsing and many more applications. Earlier mobile phones will be big and can't be carried in packet and at the same time it won't offers many applications, so a development or improvement of mobile phone is highly needed. And the development came in a very quick time.

The first mobile phone was very heavy and too big to carry, but the mobiles have a good growth that they have been now very small and they are up to feather touch. Mobile phone evolution has been very quick because of need of users, mobile development have been very much influenced by their manufactures. The major production of mobiles has been shared between Nokia, Samsung, Motorola, blackberry, apple and etc. The competition between these manufactures makes the mobile phone technology to grow very higher, so mobiles with advanced technology evolved; they were user friendly with many options and applications.

More the advanced in mobile technology more they increase in numbers. Due to its mobility, mobile usage has been increasing very much day by day. In next ten years or so the usage of mobile phones will be unavoidable and every individual in the world tries having one for them.

## Development of Mobile Phone

This can be viewed as mobile generations. There are three generations namely 1G, 2G and 3G. The first generation mobiles are analogue one, second are digital and the third generation mobiles gives us multimedia communication. ITU refers IMT 2000 as definition of 3G. The advanced researechof IMT being studied by ITU which is known as fourth generation. Half the world's 6.5 billion people now use a mobile (up from Two billion just two years ago).

The growth of the mobile also depends on the software's used in it, at present there are many OS are being employed for development of mobile. A statistics showing the market share of software's. Symbian OS had a 46.6% share of the smart mobile devices shipped in third quarter of 2008, with Apple having 17.3% (through iPhone OS), RIM having 15.3% and Microsoft having 13.5% (through Windows CE and Windows Mobile) Other competitors include Palm OS, Qualcomm's BREW, Google Android, SavaJe, Linux and MontaVista Software.

The usage of mobile increased in early 2000, because of decrease in rates of mobile and offers from the network providers and various attracting models.

## Mobile Phone Usages

At the initial stage mobiles were used only for communication purposes, but later they were improved to provide some applications like games calculator. Further on they were enhanced to support internet (browsing) and many applications which are implemented using internet. Mobile phone along with calling purposes they also used for sending the text messages which is a easier and trustful way of sending messages. SMS is also cheap when compared to calling charges, the below graph shows the SMS usage worldwide from January 2000 to May 2002. Even SMS have played a role among young customers to buy mobiles and so increase in mobile usage.

### Sms Usage World Wide

### Future Mobile Phone

Ever since from the invention of the mobile phone, development was very massive. Taking that into consideration the future of mobiles will be a tremendous one. I believe future of the mobile phone can be viewed in two ways i.e. technology and increase in no of users. Already numbers of users of mobile are in millions; in next ten years nearly every individual in developed and developing countries will have mobile phones.

### Modern Mobile Phone

In the technology side development of mobile will be unimagined, because at present mobiles are changing to touch and GPRS phones. The development will be that the mobile phones can turn into minicomputer with enhanced options and I personally believe instead of touching or pressing buttons in mobiles can be developed wholly to usage of voice signals i.e. the mobile can be operated with the help of user's voice. So mobile will be more effective and more easy to use.

Mobile phone now exits will go through a huge change in the next ten years, they will be advanced with many applications to support. In short mobile be replacing landline phones computers and be a more sophisticated smart phone.

## Wearable Computers

The terms "wearable technology", "wearable devices", and "wearables" all refer to electronic technologies or computers that are incorporated into items of clothing and accessories which can comfortably be worn on the body. These wearable devices can perform many of the same computing tasks as mobile phones and laptop computers; however, in some cases, wearable technology can outperform these hand-held devices entirely. Wearable technology tends to be more sophisticated than hand-held technology on the market today because it can provide sensory and scanning features not typically seen in mobile and laptop devices, such as biofeedback and tracking of physiological function.

Wearable computing facilitates a new form of interaction between the human and the computer comprising a small body worn computer that is always accessible and ready for use. A wearable computer used for such operations is incorporated into the personal space of the user, control by the user and is always on and accessible. For a layman, it is like becoming a 'Superhero', where you can be at more than one place at the same time. A wearable computer is more than just a wrist watch or regular eyeglass, rather it functions just like a computer system. The fact that segregates wearable devices from wearable computer is that unlike those other wearable devices that are not programmable; the wearable computer is as reconfigurable as the familiar desktop.

These computers contrary to their size can perform, if not more than equal number of computations as any PC or laptop:

Test 1 - Wearable - These computers should be wearable for extended time periods with the user experience being significantly enhanced.

Test 2 - Smart - They should have advanced circuitry, wireless connectivity and at least a minimum level of independent processing capability.

Test 3 - Should provide constant access to information services. They are designed for everyday use and should interact with the user at any given time interrupting when necessary.

Test 4 - Wearables should sense the user's internal state and should provide the best cognitive support.

Test 5 - They should take minimal necessary attention of the user and should guarantee the privacy and encourage personalization.

Test 6 - Wearables should mediate between automation or computation in the environment and the user to present a interface consistent with the user's preference and abilities.

## Features of Wearable Computer

The salient features in this novel interaction between human and computer which make it outstanding are:

- Consistency: The computer runs continuously and is user friendly. Unlike, a hand held devise like tablets or laptops, it does not need to be opened up and turned on prior to its use. The signal flows from human to computer and computer to human continuously to provide a constant user interface.
- Enhancement: Traditional computing archetypes are based on the notions that computing is the primary task. On the contrary, wearable computing does not follow the above notion. The assumption of the wearable computing is that the user will be doing something else along with computing. Thus, the computer should serve to increment the intellect or augment the senses.
- Mediation: Wearable computing unlike traditional computing acts as a mediator as it aids its users by providing its umpteen applications in medical care, domestic use, corporate world, military etc.
- Privacy: Wearable computing can be used to create a new level of privacy because it is much more personal as it is worn so it helps to develop a close synergy between human and computers.

- Convenient: Wearable technology is of utmost convenience to the user as the right person can use it at the right time and at the right place which offers a great comfort zone and helps to increase its overall utility.
- Unrestrictive: Wearable devices enable a person to do multi-tasking and do not restrict the focus of a person to one particular thing. For example to access a data on a PDA (Public Digital Assistant), the user must interrupt what he is doing and then focus on the devise. This restricts the activities of the user unlike in the case of wearable computing.

## Application

The implication and uses of wearable technology have a widespread impact and influence the fields of medical care, fitness, aging, disabilities, education, transportation, enterprise, finance, gaming and music. The final aim of wearable technologies in each of these fields will be to efficiently incorporate functional, portable electronic and computers into individuals' daily lives. Prior to their presence in the consumer market, wearable devices were primarily used in the field of military technology and had the biggest implications for healthcare and medicine.

Table: Applications of wearable computing.

| Application | Product Categories | |
|---|---|---|
| Healthcare And Medical | Blood pressure Monitors Continuous Glucose Monitoring<br>Defibrillators Drug Delivery Product ECG Monitoring Hearing Aids | Insulin pumps<br>Smart Glasses<br>Patches<br>PERS<br>Pulse Oximetry |
| Fitness and Wellness | Activity Monitors Emotional Measurement Fitness &Heart rate Monitors Food Pods & Pedometers Heads-up Displays | Sleep Sensors<br>Smart Glasses<br>Smart Clothing<br>Smart Watches Other, Audio Ear buds |
| Infotainment | Bluetooth Headsets Head–up Displays Imaging Products Smart Glasses Smart Watches | |
| Military | Hand-Worn Terminals<br>Head-up Display<br>Smart Clothing | |
| Industrial | Hand-Worn Terminals<br>Head-up Display<br>Smart Clothing<br>Smart Glasses | |

## Augmented Memory

The Remembrance Agents (RAs) are wearable that continuously reminds the wearer of potentially relevant information based on the wearer's current physical and virtual context. RAs are always on and always active and working, instead of being 'worker up' when needed.

## Finger Tracking

This is one of the simplest applications of camera based wearable computing. The computer would be able to visually track the user's finger. The user would then be able to control the computer with his fingers as well as he would with a mouse.

### Face Recognition

Working together with an appropriate face finding software, face recognition system can be adapted for use in wearable computing. Market users include the police, politicians, teachers, the visually challenged and those with bad memories.

### Visual Filter

This is particularly useful for the visually disabled. The wearable computer can digitally magnify an image or a prose through the use of a virtual fish eye lens to assist in reading. This can be done through the use of a digital visual filter.

### Navigation

Navigation 8 connecting a Global Positioning System (GPS) to wearable and certain mapping software allows the users to track their desired or the current location while exploring a new city. A visually impaired person might be able to receive warnings of approaching object and hence promote safety in their daily lives.

### Repair Instruction

By putting as little as 3 exclusive marks at fixed distances from each other, a wearable camera with known focal length can recover the 3-D location of an object defined by these three marks. The wearable automatically determines the hazard, locates the 3D position of the object and prepares specific 3D real time step by step guidelines on the object for the technician to follow.

### Communication Management

A WAC can be used to manage personal communications naturally with much mobility. Speech recognition can be used to convert the user's responses (with constraints on vocabulary and grammar) into text for email responses. This, in turn, combines speech recognition, synthesis and digital audio recording to act as a virtual secretary which manages mobile communications.

### Remote Sensing and Maintenance

Wearable can provide field workers by giving them remote assistance and expertise through digital data, audio and image. With this, even non expert maintenance personal can accomplish simple repairing tasks with aid of remote experts at the help desk.

### Industrial

The quick and accurate availability of complex information to the workers in the field, office workplace or in a non-office workplace has been important to many organizations since the establishment of computerized records in the 1950s. Many workers need to use one or both hands while carrying out their tasks, and also need to maintain eye contact with taking into account the protection aspect. Depending upon the type of industry, wearable computers have varied advantages. Wearable Computers, because of their flexibility, portability and hands free use, are being used by workers in many types of industries. The wearable computer with a hands free interface, e.g. speech, and a head-mounted display can provide a solution for these workers. The first organization to recognize this, and to commit resources to researching possibilities, was The Boeing Company. In the 1990's Boeing employed several hundred staff to assemble wiring harnesses for aircraft. These wire bundles were constructed using pegs in a number of 3' x 8' easel-like form boards with paper printouts glued to their surfaces, and a separate set of printouts for reference. It was proposed that a worker with a head-mounted augmented reality display could be guided through the assembly task with no need for reference to paper print-outs. It demonstrated that such a system was practical and identified limitations which would be addressed as technology improves. Worker issues such as safety, comfort and social compatibility needed special consideration.

Another example is the Mercedes-Benz pebble smart watch. The idea is that drivers use the sleekly styled smart watch to check on their vehicle's location, door lock status and even fuel level. The nifty watch can alert its wearer of incoming calls, texts and emails, so too can it in theory warn the driver of real-time hazards when behind the wheel.

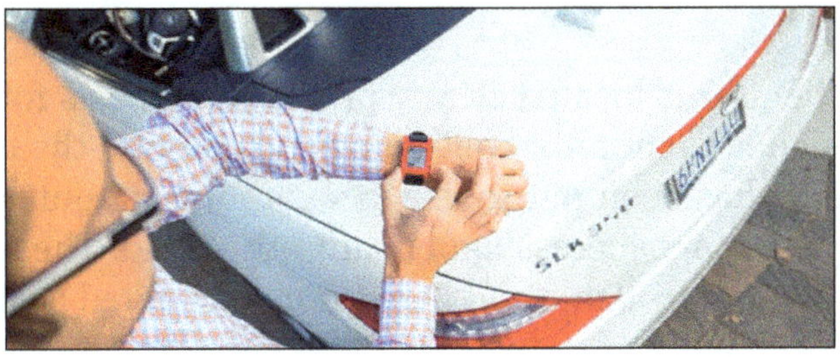

Figure: Mercedes-Benz pebble smart watch.

## Military

The potential applications of wearable computers to an infantryman were quickly recognized by organizations and other law agencies.

Apart from providing command/control communication and navigation functions, a wearable could give access to tactical information that could help distinguishing between friendly and hostile forces, and potentially offering strategies for dealing with dangerous scenarios. Wearable computers will help infantryman use digitally distributed orders, maps, and intelligence. Wireless technology enables each soldier to communicate seamlessly with his comrades within a networked unit.

Eventually soldiers would have the capability to exchange video of what they see with each other via miniature cameras interfaced with lightweight wearable computers. Wearable for military would have to be more rugged and robust to withstand knocks and shocks.

Many military wearable computers are designed with the total integration of man and system in mind. Much of the research of wearable in military has been kept confidential however examples of collaboration with non-military researchers can be found in the United States, Australia, the United Kingdom and Singapore.

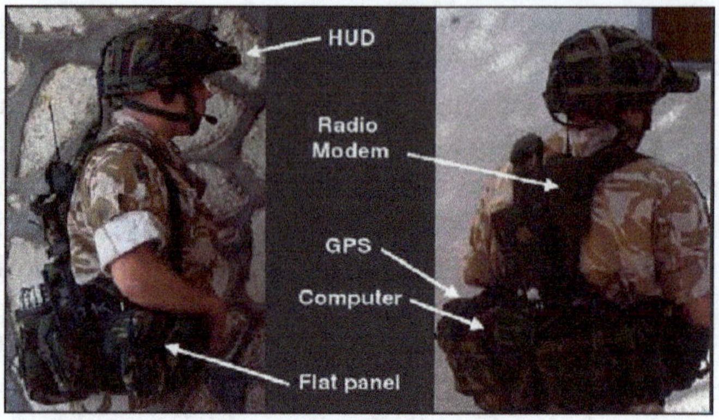


Figure: Military wearable.

The 9Solutions IPCS is a wireless Bluetooth and SaaS-based real-time locating system (RTLS) and provides a platform which enables real-time identification of people and equipment. The U.S. Army in particular have funded the Land Warrior program which initially provided positioning and targeting information, battlefield communications, and thermal sight imaging from the soldier's weapon.

## Medical

The knowledge of where the user is located clearly provides the basis for many wearable designs. Wearables can also be designed to monitor well-being and activity -the how and what of the user. This form of context sensing has been put to use in wearable computers for medical and health applications and has met with more success than in any other field. The most common is the pacemaker. However it is not user controllable and hence not considered a wearable computer.

Examples of treatment by a wearable are insulin pump therapy for diabetics and a brain implant to facilitate communication with speech-incapable patients.

Figure: Health monitoring.

More recently health monitoring wearables have become commercially available in the form of the Body media product range (Bodymedia). They are based around an armband design with sensors for detecting movement, heat flux, skin temperature, near-body temperature, and galvanic skin response. Data can be either viewed in real time via a wireless link, or downloaded for analysis using the Internet.

Many systems have been explored to provide the visually impaired with guidance. Early examples of this were developed at the University of California, Santa Barbara using GPS.

The Metria Wearable Sensor is an entirely new tool aimed at assisting the healthcare industry shift toward prevention. The user attaches the wearable sensor, which uses "skin-friendly" cohesive; the sensor collects data, such as the breaths per minute and number of hours slept and breaths per minute. The Remote Monitoring System (RMS) system was developed in collaboration with Mayo Clinic and uses designs to support remote monitoring for patients with cardiac problems. Examples include the Electroencephalography (EEG), BodyTel Products and BioMan t-shirt.

## Sports and Fitness

Interest in wearables has so far been largely driven by consumer interest in sports and fitness. The typical functions of such wearables are performance monitoring, activity tracking, goal monitoring and management, direction data and location share and optimum performance management.

The BioHarness is one such wearable and is an exclusive compact analysis of monitoring module that enables the transmission and capture of physiological data on the wearer via mobile and fixed data networks enabling genuine remote monitoring of human performance and condition in the real world.

Another example are the Smart Training Shoes which are the next generation of athletic shoes will feature radio frequency identification tags, motion sensors and accelerometers that will allow you to customize the look, fit and responsiveness of your kicks.

Also an activity tracker is a device or application for monitoring and tracking fitness-related metrics such as distance walked or run, calorie consumption, and in some cases heartbeat and quality of sleep.

Figure: Nike Smart Training Shoes.

## Tourism

A few wearable computers which are promising for tourists and might prove to be a boom for the tourism industry are given below:

## Narrative Clip

It is a small wearable life logging camera that pins to the chest (or wherever). It records 2 photos a minute. Users can set this to a faster setting if they like, providing a clear record of life lived.

## Tourist Diary

Tourist diary is to help a tourist organize his trip experience by multimodal interaction. A tourist can request the system to take picture or video clip which he can add or dictate to his diary

Tourist Diary is to help him organize his/her diary. After the tourist if finished the site he could ask the system to generate an html document based on the stored info which he can then share with others.

### Triposo Travel Belt

To help a tourist find their destinations without having to look in their phone every time, Triposo is making a belt called the Travel Belt. It feeds explorers, with up-to-date information, maps, intelligent recommendations, tips and tricks, etc.

## Advantages

### Consistent

There is always a constant interaction between the computer and the user and hence there is no need to turn it on or off.

### Multi-Tasking

Wearable computer provides computational support even when the user's hands, voice, eyes, or attention is actively engaged with the physical environment. Other than its very efficient partaking in recognizing a person in a high alerted area, it can be manipulated to complete daily tasks of the wearer tracking his/her daily habits.

### Mobile

Wearable computers must go where the wearer goes. They are always on and their wearer can access the many time. Adding to our convenience, wearable computing enables a person to work from anywhere, literally.

### Unrestrictive

When using it you can do other things.

### Communicative

It can be used as a communications medium when you want it to. Any person in possession of these devices can communicate with their enterprise information systems without suspending his/her routine work.

### Can't be Lost Easily

A personal wearable device adheres to the wearer's needs and is very unlikely to get lost also it being embedded to one's clothes as opposed to the handheld devices.

### Attentive

Wearable computer is environmentally aware, multimodal, and multisensory. These devices are relentlessly acting out contextual tasks of providing prompts or reminders, apprehending experiences or information and salvaging time-sensitive information in support of the user.

### No Distraction

Wearable computer does not cut you off from the outside world such as a virtual reality game or the like.

## Disadvantages

### Expensive

Wearable computers are one of the most cultured pieces of technology existing today. Sophistication comes with expense as it requires high levels of technology to come up with wearable computers making the end product high priced and extravagant to many. In the case of a construction company, wearable computers have the potential to increase overall efficacy but are undermined by the high costs of setting up local area networks, LANS, which would aid in the synchronization of data.

### Heavy

Wearable computers can be quite heavy. This is due to the many components of the wearable technology that has to be attached to the body. A wearable computer, like any other computer, requires a Central Processing unit (CPU) as well as peripheral devices and a monitor that enable the computer to gasp the input of data. These components are hefty and difficult to wear so it causes enormous amount of problem to the user. The computer can also be quite heavy if all these components are built into the wearable computer.

### Discomfort

Wearable computers tend to irritate the user at the time of hot and humid weather conditions as these wearable components emit heat despite the cooling technology inbuilt in them. It also results in health hazards such as headache and dizziness.

### Security

If left unattended, wearable technology can cause several security issues as it can be hacked easily. Wearable computers are connected to the company's server to enable easy flow of communication between people out in the field and their counterparts in the office. However, if not tackled carefully, then the outer source can use this information to either steal company or personal secrets.

## Personal Digital Assistant

A personal digital assistant (PDA) is a portable device that functions as a personal information manager. PDAs are used for Web browsing, office applications, watching videos, viewing photos or as mobile phones.

The main purpose of a personal digital assistant (PDA) is to act as an electronic organizer or day planner that is portable, easy to use and capable of sharing information with your PC. It's supposed to be an extension of the PC, not a replacement.

PDAs, also called handhelds or palmtops, have definitely evolved over the years. Not only can they manage your personal information, such as contacts, appointments, and to-do lists, today's devices can also connect to the Internet, act as global positioning system (GPS) devices, and run multimedia software. What's more, manufacturers have combined PDAs with cell phones, multimedia players and other electronic gadgetry. As its capabilities continue to grow, the standard PDA device is changing.

Predecessors of the modern PDA include the Psion Organiser and the Sharp Wizard. These early devices, which were intended to be portable computers, came out in the mid- to late 1980s. They included small keyboards for input, a small display, and basic features such as an alarm clock, calendar, telephone pad and calculator. Support for specialized software such as games and spreadsheets was also included. The Psion Organiser II, released in 1986, was especially popular and more than half a million were sold.

In 1993, Apple introduced the Newton MessagePad at a price of $700. It provided users with an electronic notepad, to-do list, calendar, telephone log and address file applications. Some of the Newton's innovations have become standard PDA features, including a pressure-sensitive display with stylus, handwriting recognition capabilities, an infrared port and an expansion slot. However, the Newton MessagePad was too big, expensive and complicated, and its handwriting recognition program was poor. Apple discontinued the Newton in 1998.

The original PalmPilot was introduced in March of 1996 by Palm Computing (owned by U.S. Robotics at the time). It cost less than $300, ran on its own Palm operating system, fit in a shirt pocket and synchronized with consumers' PCs. The PalmPilot ran for weeks on AAA batteries, was easy to use, and could store thousands of contacts, appointments and notes. Part of its small size was due to the lack of a keyboard. Users used a stylus and the Graffiti language to input data.

Not to be outdone, Microsoft had been working on various forms of portable computing, including PenWindows and tablet computers. In November 1996, Microsoft released Windows CE, its first operating system for mobile devices. A number of manufacturers, such as HP, Compaq and Casio, adopted it for what was dubbed the Handheld PC -- the first Windows-based competitor for the PalmPilot.

## Types of PDAs

## Traditional PDAs

Today's traditional PDAs are descendents of the original PalmPilot and Microsoft Handheld PC devices. Palm devices run the Palm OS (operating system), and Microsoft Pocket PCs run Windows Mobile. The differences between the two systems are fewer than in the past.

## Palm PDAs

Most Palm devices are made by palmOne, which offers the Zire and Tungsten product lines. The company formed in 2003 when Palm Computing acquired Handspring, Inc. Sony, which produced the Palm-based CLIE, stopped producing PDAs in 2005.

Known for their ease of use, Palm OS PDAs have:

- A vast library of third-party applications (more than 20,000) that you can add to the system. (most devices come bundled with e-mail, productivity and multimedia software).
- An updated version of the Graffiti handwriting-recognition application.
- Synchronization with both Windows and Macintosh computers using the Palm Desktop.
- Smaller displays than Pocket PCs to accommodate a dedicated Graffiti area on the device. (Some higher-end Palm devices now incorporate a virtual Graffiti area in the display, resulting in a larger display area).

## Pocket PCs

Pocket PC is the generic name for Windows Mobile PDAs. Their standard features include:

- Pocket versions of Microsoft applications such as Microsoft Word, Excel, and Outlook. (note that some formatting is lost between Pocket and standard versions of documents).
- Synchronization with Microsoft Outlook on a Windows PC.(synchronization with e-mail systems other than Outlook or with Macintosh computers requires additional software).
- Three handwriting-recognition applications: Transcriber, Letter Recognizer (similar to the new version of Graffiti), and Block Recognizer (similar to the original Graffiti).
- A virtual writing area, which maximizes the display size.
- Windows Media Player for multimedia content.

## Smartphones

A smartphone is either a cell phone with PDA capabilities or a traditional PDA with added cell phone capabilities, depending on the form factor (style) and manufacturer. Characteristics of these devices include:

- A cellular service provider to handle phone service. (As with cell phones, you typically purchase a cellular plan and smart phone from the service provider).
- Internet access through cellular data networks.

- Various combinations of cell phone and PDA features, depending on the device (for example, not all smart phones offer handwriting-recognition capabilities).
- A number of different operating systems, including Windows Mobile Pocket PC Phone Edition, the Palm OS, the Blackberry OS for Blackberry smart phones, and the Symbian OS for smart phones from Panasonic, Nokia, Samsung and others.

Related devices include ultraportable computers, smart watches and multimedia players. Ultraportable computers range in weight from less than a pound to about three pounds, come with a full desktop operating system (such as Windows XP), and include a small keyboard. Smart watches offer some PDA functions in a wristwatch form factor. Microsoft's SPOT (Smart Personal Objects Technology)-based watches, for example, can receive MSN Direct information such as weather and news. They can also receive calendar information and personal messages. Some multimedia players can combine the functions of a PDA with multimedia features, such as a digital camera, an MP3 player and a video player.

### PDA Features

Even the most basic PDAs handle standard personal information management (PIM) functions, run application software and synchronize with PCs.

### Handle Standard PIM Functions

All PDAs come with some kind of personal information management (PIM) software that typically handles the following tasks to keep you organized:

- Store contact information (names, addresses, phone numbers, e-mail addresses).

- Make to-do lists.
- Take notes.
- Track appointments (date book, calendar).
- Remind you of appointments (clock, alarm functions).
- Perform calculations.

## Run Application Software

PDAs can run specialized software applications:

- Windows Mobile devices come with Pocket versions of Word, Excel, Internet Explorer and Outlook (includes e-mail and PIM functions), along with Windows Media Player and voice memo recording.
- Most Palm OS devices include applications such as DataViz Documents to Go (compatible with Microsoft Word, Excel, and PowerPoint), palmOne Media (for photos and video), VersaMail e-mail software and web-browsing software.
- All types of devices can run other kinds of software including games, multimedia, expense, diet and exercise, travel, medical, time and billing, and reference.

## Synchronize with PCs

Because PDAs are designed to complement your PC, they need to work with the same information in both places. If you make an appointment on your desktop computer, you need to transfer it to your PDA; if you jot down a phone number on your PDA, you should upload it later to your PC.

Synchronization software on the PDA works with companion software that you install on your PC. Microsoft Pocket PC devices use ActiveSync and Palm OS devices use HotSync synchronization software. On your computer, you also need an application like Microsoft Outlook or the Palm Desktop that holds PIM information on the PC side. The beauty of synchronization is that you always have a backup copy of your data, which can be a lifesaver if your PDA is broken, stolen, or completely out of power.

## In Sync

To synchronize (also called sync) a PDA:

- Connect your PDA to a PC with a cable or cradle/cable combination, normally through USB ports or serial ports located on the PDA and your PC. Some devices also support synchronization and data transfer through wireless connections such as Bluetooth.
- Push the sync button (on the device or cradle) to start the synchronization process.
- Wait as the process compares the files and information between the PDA and PC, and then copies the most current versions of the files to both the PDA and the PC.

## Common PDA Functions

Today, most PDAs incorporate wireless and multimedia functions of some type. Functions found on most (but not necessarily all) devices include:

- Short-range wireless connectivity using Infrared (IR) or Bluetooth technology, IR is found on most PDAs and requires a clear line of sight. It's commonly used to sync with a notebook computer that has an IR port. Bluetooth wirelessly connects (it's a radio frequency technology that doesn't require a clear line of sight) to other Bluetooth-enabled devices, such as a headset or a printer.
- Internet and corporate network connectivity through Wi-Fi and wireless access points.
- Support for Wireless WAN (Wide Area Networks); the cellular data networks that provide Internet connectivity for smart phone devices.
- A memory card slot that accepts flash media such as CompactFlash, MultiMediaCard, and Secure Digital cards (Media cards act as additional storage for files and applications).
- Audio support for MP3 files and a microphone, speaker jack and headphone jack.

Figure: SanDisk 256 MB Secure Digital Card.

## Bells and Whistles

High-end PDAs offer multimedia, security and add-on features not found on less expensive devices:

- A Secure Digital Input/Output (SDIO) card slot for add-on peripherals contained in an SDIO card, for example, a Bluetooth card, a Wi-Fi card, or a GPS (global positioning system) card.
- Built-in GPS capabilities.
- A built-in digital camera for snapping digital images and capturing short videos (The quality will not be as good as that of a dedicated camera).
- Integrated security features such as a biometric fingerprint reader.

## The PDA Computer

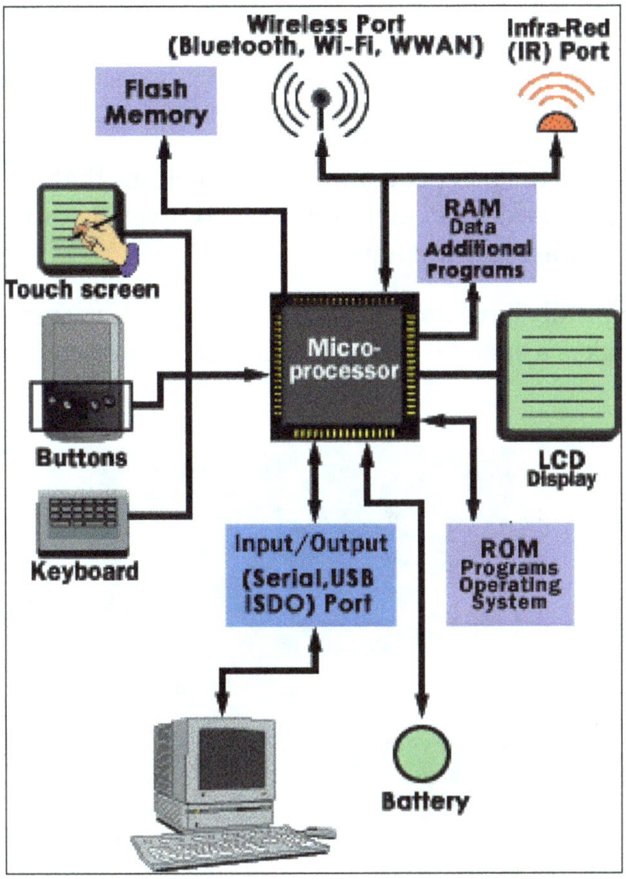


Figure: The parts that can make up a PDA.

## Microprocessors and Memory

Like standard desktop and laptop computers, PDAs are powered by microprocessors. The microprocessor is the brain of the PDA, and it coordinates all of the functions according to programmed instructions. Unlike desktop PCs and laptops, PDAs use smaller, cheaper microprocessors. Although these microprocessors tend to be slower than their PC counterparts, they're adequate for the tasks that PDAs perform. The benefits of small size and price outweigh the cost of slow speeds.

A PDA doesn't have a hard disk. It stores basic programs (address book, calendar, memo pad and operating system) in a read-only memory (ROM) chip, which remains intact even when the machine shuts down. Your data and any programs you add later are stored in the device's random-access memory (RAM). Information in RAM is only available when the device is on. Due to their design, PDAs keep data in RAM safe because they continue to draw a small amount of power from the batteries even when you turn the device off.

Less powerful PDAs have lower amounts of RAM. However, many application programs take up significant memory space, so most models have more memory. Also, Pocket PC devices generally require more resources and have even more RAM. To provide additional memory, many PDAs accept removable flash media add-on cards. These are handy for storing large files or multimedia content, such as digital photos.

Some newer PDAs, such as the Palm Tungsten E2, use flash memory instead of RAM. Flash memory is non-volatile, which means it preserves the data and applications it stores - even when all battery power is depleted.

## Operating Systems

The operating system contains the pre-programmed instructions that tell the microprocessor what to do. The operating systems used by PDAs are not as complex as those used by PCs. They have fewer instructions, which require less memory.

PDAs and smartphones typically have one of two types of operating systems: Palm OS or Windows Mobile. However, RIM makes a specific OS for its BlackBerry devices, and the Symbian OS operates some smartphones.

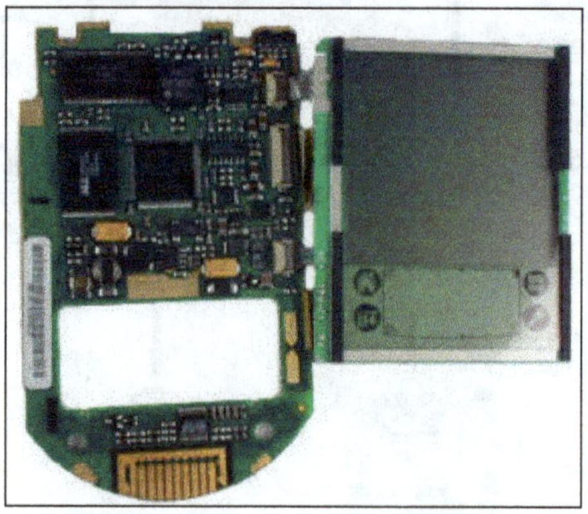

Figure: Here's an inside view of a PDA. The circuit board folds away from the screen. In the middle of the single-layer circuit board is the microprocessor, and to the left and above are the memory chips.

## PDA Battery

So far, we've looked at the PDA's microprocessor, memory and operating system. Now let's look at the other parts of a PDA.

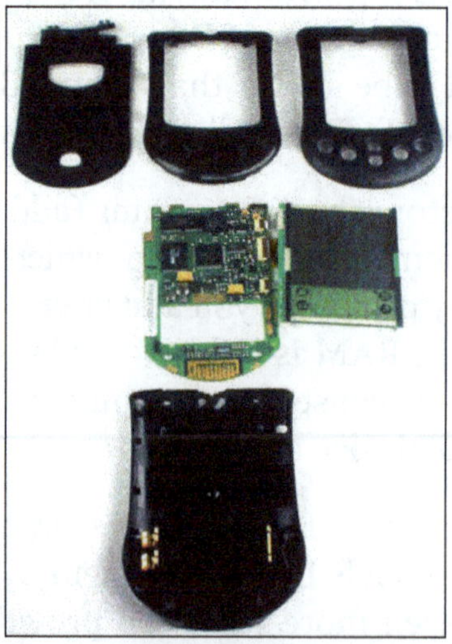

Figure: Here are the parts of the PDA -- the case, the LCD screen and the circuit board. This model comes in basic black, but you can buy interchangeable covers in various colors.

## Batteries

PDAs are powered by batteries. Some models use alkaline (AAA) batteries, while others use rechargeable batteries (lithium, nickel-cadmium or nickel-metal hydride). The battery life depends on what kind of PDA you have and how you use it. Here are some of the things that can drain batteries:

- Operating system (PocketPC requires more power by virtue of its increased memory requirements).
- More memory.
- Wireless connections, such as Wi-Fi and Bluetooth.
- Backlighting on the display.

Battery life can vary from hours to months, depending upon the PDA model and its features. Most PDAs have power management systems in place to extend the battery life. Even if the batteries are so low that you can no longer turn the machine on (it will give you plenty of warning before this happens), there's usually enough power to keep the RAM refreshed.

If the batteries do run completely out of juice or if you remove them, most devices have an internal backup battery that provides short-term power (typically 30 minutes or less) until you install a replacement. If all power sources are depleted, PDAs lose all of the data in RAM. This makes backing up or synchronizing your PDA extremely important. In addition to battery power, many PDAs come with AC adapters to run off household electric currents. A car adapter is also generally available as an accessory.

## LCD Display

PDAs use an LCD (liquid-crystal display) screen. Unlike the LCD screens for desktop or laptop computers, which are used solely as output devices, PDAs use their screens for output and input. The LCD screens of PDAs are smaller than laptop screens, but vary in size. Almost all PDAs now offer color displays.

PDA displays have the following features:

- Transflective TFT (thin-film transistor) LCD for indoor and outdoor use.
- Different pixel resolutions with higher resolutions for better quality.
- Color screen.
- Backlighting for reading in low light.

## Input Methods

PDAs vary in how you input data and commands. Some devices use a stylus and touch screen exclusively in combination with a handwriting recognition program. Using a plastic stylus, you draw characters on the device's display or dedicated writing area. Software inside the PDA converts the characters to letters and numbers. On Palm devices, the software that recognizes these letters is called Graffiti. Graffiti requires that each letter be recorded in a certain way, and you must use a specialized alphabet. For example, to write the letter "A," you draw an upside-down V. The letter "F" looks like an inverted L. To help Graffiti make more accurate guesses, you must draw letters on one part of the screen and numbers in another part.

Pocket PC PDAs offer three handwriting-recognition applications: Transcriber, Letter Recognizer and Block Recognizer. Letter Recognizer and Block Recognizer are similar to Graffiti and require specialized alphabets. By contrast, Transcriber recognizes your "regular" handwriting, as long as you write legibly. It is similar to the handwriting recognition capabilities found on Tablet PCs.

If you can't get the hang of PDA handwriting, you can use a miniature onscreen keyboard. It looks just like a regular keyboard, except you tap on the letters with the stylus. In addition, many devices now include a small (and usually cramped) QWERTY keyboard. Some of these require you to use your thumbs to type. And you can use a full-size keyboard by connecting it to the PDA via Bluetooth or a USB port. Each model also has a few buttons and navigation dials to bring up applications and scroll through files.

## PDA Display and Input

### LCD Display

PDAs use an LCD (liquid-crystal display) screen. Unlike the LCD screens for desktop or laptop computers, which are used solely as output devices, PDAs use their screens for output and input. The LCD screens of PDAs are smaller than laptop screens, but vary in size. Almost all PDAs now offer color displays.

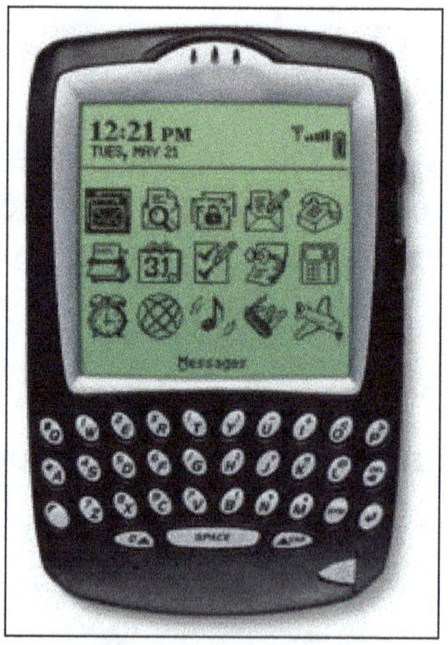


PDA displays have the following features:

- Transflective TFT (thin-film transistor) LCD for indoor and outdoor use.
- Different pixel resolutions with higher resolutions for better quality.
- Color screen.
- Backlighting for reading in low light.

## Input Methods

PDAs vary in how you input data and commands. Some devices use a stylus and touch screen exclusively in combination with a handwriting recognition program. Using a plastic stylus, you draw characters on the device's display or dedicated writing area. Software inside the PDA converts the characters to letters and numbers. On Palm devices, the software that recognizes these letters is called Graffiti. Graffiti requires that each letter be recorded in a certain way, and you must use a specialized alphabet. For example, to write the letter "A," you draw an upside-down V. The letter "F" looks like an inverted L. To help Graffiti make more accurate guesses, you must draw letters on one part of the screen and numbers in another part.

Pocket PC PDAs offer three handwriting-recognition applications: Transcriber, Letter Recognizer and Block Recognizer. Letter Recognizer and Block Recognizer are similar to Graffiti and require specialized alphabets. By contrast, Transcriber recognizes your "regular" handwriting, as long as you write legibly. It is similar to the handwriting recognition capabilities found on Tablet PCs.

If you can't get the hang of PDA handwriting, you can use a miniature onscreen keyboard. It looks just like a regular keyboard, except you tap on the letters with the stylus. In addition, many devices now include a small (and usually cramped) QWERTY keyboard. Some of these require you to use your thumbs to type. And you can use a full-size keyboard by connecting it to the PDA via Bluetooth or a USB port. Each model also has a few buttons and navigation dials to bring up applications and scroll through files.

## Buying a PDA

If you're in the market for a PDA, the main question you should consider is, "What will I be using this for most of the time?" If you're looking for basic PIM functions and organization, you probably don't need wireless connectivity. On the other hand, if you travel frequently and want to stay in touch via e-mail, you need Wi-Fi capability or a smartphone with cellular data service.

Because most people use a PDA as a PC accessory, synchronization is an important consideration. Look for a device that easily synchronizes with the PIM software you use on your computer. For example, if you have a Macintosh computer or you don't use Microsoft Outlook, a Palm OS device may be more to your liking.

The device's data entry method is another important consideration. If you plan to use your PDA for e-mail or other text-heavy applications, consider a built-in keyboard. If you can, try out the keyboard before you buy.

The size and shape of the device and its display are also important considerations. The device should be comfortable to hold. The display size and clarity should also meet your expectations. As with the keyboard, it's best to make a trip to a store that sells the devices you're interested in so you can try it out before you buy.

If you are interested in a multipurpose device that combines features such as GPS, an MP3 player, or digital camera, look for a device that has these features integrated. Alternatively, you can opt for a device with an SDIO slot and use compatible SDIO peripherals. If you plan to use numerous applications or store large files, look for a device that accepts flash memory media cards.

Accessories to consider include an additional battery, a car or travel charger, headphones, a carrying case, a plastic screen protector and add-on keyboards.

If you're interested in a smartphone, many of the considerations for purchasing a cell phone apply. Some of the most important things to check include service providers' coverage area, the additional cost for cellular data service (if any), and the length of the service contract.

## E-Reader

An e-reader, or "e-book reader," is a portable hardware device designed for reading digital publications. These include e-books, electronic magazines, and digital versions of newspapers. Since textual data does not require a lot of storage space, most e-readers can store thousands of books and other publications. Just like an iPod can store an entire music library, a single e-reader can store a large collection of books.

Dozens of different e-readers are available, but some of the most popular ones include the Amazon Kindle, the Barnes and Noble Nook, and the Sony Reader. These devices all support a wide range of eBook formats and can download content over a wireless network. Many e-readers have a monochrome display, often called "electronic paper," while others have a full-color backlit display. While the electronic paper displays do not provide color images, the screen appears more like a paper page from a book, and it can be easily viewed in bright sunlight.

Tablets, such as the Apple iPad, the BlackBerry PlayBook, and the Amazon Kindle Fire are often considered e-readers, since they can be used for reading digital publications. However, it is more accurate to refer to these devices as tablets that can be used as e-readers since they are not designed primarily as digital readers. Tablets offer more capabilities than e-readers, but e-readers are often better suited for just reading e-books.

## One Device, Many Books

eBooks are portable and lightweight, making it easy to carry around. Instead of carrying multiple bulky books, one eBook reader can hold thousands of eBooks. It saves a lot of space- in your home and in your bag. One doesn't have to worry about the storage limit.

A single device is enough to read any number of books you want. Students benefit the most out of it as they do not have to carry a bag full of books every day. It allows the user to store plenty of books on the device, way more than anyone can read in their lifetime.

## Accessible Everywhere

eBooks can be downloaded and stored for later use. One can carry the eBook around and read them whenever they want. Students and employees can go through the learning material while at home and even while travelling. It is convenient for people on the go. Some eReaders come with the option of offline accessibility, allowing people to use it even in the absence of an internet connection.

Unlike printed books which have to be ordered or purchased from the bookstore, with time lost in waiting for delivery or travelling to the bookstore; readers can easily access any eBook anytime and anywhere.

## Easily Updates

The contents on the eBooks are mostly cloud-based, meaning it can be updated any time. Authors and publishers can always add the up-to-date information, providing users with the latest digital content. This saves reprinting costs and the time involved in the process.

## Shareable Content

You can share the eBook contents with multiple users. The social feature on the ebook allows sharing and liking of content, which is not possible with printed books.

You might share a printed book with one person at a time, but an eBook can be shared with many people at once. Students and employees can use this feature to collaborate with their peers.

## Augmented Reality Experience

Augmented reality has taken the eBook experience to a whole new level. Texts and images can now be embedded with an augmented three-dimensional visual which pops up on the user's screen. Readers can view a computer-generated 3D model of the image, which makes reading and learning a much more immersive experience.

The user only has to point the device at the image or text, and an enhanced version of the image will be displayed on their screen. Many educational institutes and organizations are adopting this technology into their eBooks to enhance user experience.

## Easy on the Eyes

eReaders these days come with features where you can adjust the brightness of the screen according to the time of the day and depending upon your preferences.

Many eReaders even have read in the dark feature, with lights that won't strain your eyes, and neither would they disturb others around you. Users can even change the fonts of the text. They can increase or decrease the size of the font. Basically, eBooks ensure that your eyes are taken care of.

## Read Aloud Feature

eBooks have been designed with text to speech or read aloud features. This allows people to listen to the eBooks if they are too busy to read it, for example, while driving, or while doing some chores.

This feature is also helpful for children and adults with a learning disability or with visual impairment. It is also useful to help the user understand how a particular word is pronounced.

### Interactive Elements

Including interactive features in an eBook makes the reading experience a more engaging one. eBooks can be embedded with audio and videos. External links can be included to offer additional information.

The user can even bookmark a certain page or find a word in a quick and effortless way. It also allows users to make annotations. There are multiple such interactive features which can be included in order to enhance the overall reading experience.

## References


- Mobile-computing: javatpoint.com, Retrieved 08, March 2020
- What-is-mobile-computing-advantages-disadvantages-applications: digitalthinkerhelp.com, Retrieved 17, May 2020
- Smart-card-technology: plugandplaytechcenter.com, Retrieved 10, January 2020
- The-development-of-the-mobile-phone-english-language-essay, english language: ukessays.com, Retrieved 03, April 2020
- Wearable-Computing-and-its-Application-306322945: researchgate.net, Retrieved 21, July 2020
- Top-10-advantages-of-ebooks-over-printed-books: kitaboo.com, Retrieved 29, May 2020

2

# Fundamentals of Mobile Communication

Mobile communication refers to the technology that allows us to communicate with others in an easier way without the need of a physical connection. High capacity load balancing, scalability, network management, access controls, roaming, modulation, antennas and propagation are some features of mobile communications which are discussed comprehensively in this chapter.

Mobile Communication is the use of technology that allows us to communicate with others in different locations without the use of any physical connection (wires or cables). Mobile communication makes our life easier, and it saves time and effort.

A mobile phone (also called mobile cellular network, cell phone or hand phone) is an example of mobile communication (wireless communication). It is an electric device used for full duplex two way radio telecommunication over a cellular network of base stations known as cell site.

## Features of Mobile Communication

The following are the features of mobile communication:

### High Capacity Load Balancing

Each wired or wireless infrastructure must incorporate high capacity load balancing. High capacity load balancing means, when one access point is overloaded, the system will actively shift users from one access point to another depending on the capacity which is available.

### Scalability

The growth in popularity of new wireless devices continuously increasing day by day. The wireless networks have the ability to start small if necessary, but expand in terms of coverage and capacity as needed - without having to overhaul or build an entirely new network.

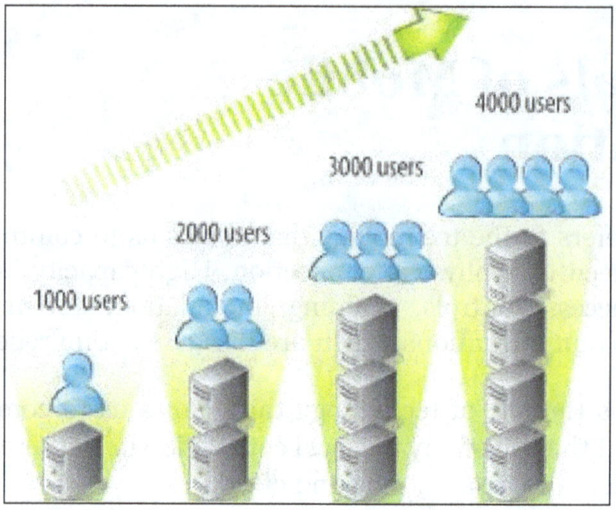


### Network Management System

Now a day, wireless networks are much more complex and may consist of hundreds or even thousands of access points, firewalls, switches, managed power and various other components. The wireless networks have a smarter way of managing the entire network from a centralized point.

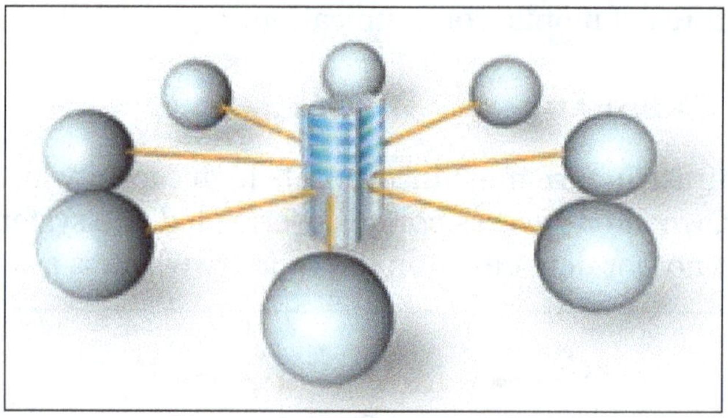

- Role based access control: Role based access control (RBAC) allows you to assign roles based on what, who, where, when and how a user or device is trying to access your network.

Once the end user or role of the devices is defined, access control policies or rules can be enforced.

### Indoor as well as Outdoor Coverage Options

It is important that your wireless system has the capability of adding indoor coverage as well as outdoor coverage.

### Network Access Control

Network access control can also be called as mobile device registration. It is essential to have a secure registration. Network access control (NAC) controls the role of the user and enforces policies.

NAC can allow your users to register themselves to the network. It is a helpful feature that enhances the user experience.

## Mobile Device Management

Suppose, many mobile devices are accessing your wireless network; now think about the thousands of applications are running on those mobile devices. How do you plan on managing all of these devices and their applications, especially as devices come and go from your business?

Mobile device management can provide control of how you will manage access to programs and applications. Even you can remotely wipe the device if it is lost or stolen.

## Roaming

You don't need to worry about dropped connections, slower speeds or any disruption in service as you move throughout your office or even from building to building wireless needs to be mobile first.

Roaming allows your end-users to successfully move from one access point to another without ever noticing a dip in a performance. For example, allowing a student to check their mail as they walk from one class to the next.

## Redundancy

The level or amount of redundancy your wireless system requires depends on your specific environment and needs. A hospital environment will need a higher level of redundancy than a coffee shop. However, at the end of the day, they both need to have a backup plan in place.

## Proper Security Means using the Right Firewall

The backbone of the system is your network firewall. With the right firewall in place you will be able to:

- See and control both your applications and end users.

- Create the right balance between security and performance.
- Reduce the complexity with:
    - Antivirus protection.
    - Deep Packet Inspection (DPI).
    - Application filtering.
- Protect your network and end users against known and unknown threads including:
    - Zero- day.
    - Encrypted malware.
    - Ransomware.
    - Malicious botnets.

### Switching

Basically, a network switch is the traffic cop of your wireless network which making sure that everyone and every device gets to where they need to go. Switching is an essential part of every fast, secure wireless network for several reasons:

- It helps the traffic on your network flow more efficiently.
- It minimizes unnecessary traffic.
- It creates a better user experience by ensuring your traffic is going to the right places.
- Advantages of Mobile Communication.

There are following advantages of mobile communication:

### Flexibility

Wireless communication enables the people to communicate with each other regardless of location. There is no need to be in an office or some telephone booth in order to pass and receive messages.

### Cost Effectiveness

In wireless communication, there is no need of any physical infrastructure (Wires or cables) or maintenance practice. Hence, the cost is reduced.

### Speed

Improvements can also be seen in speed. The network connectivity or the accessibility was much improved in accuracy and speed.

### Accessibility

With the help of wireless technology easy accessibility to the remote areas is possible. For example, in rural areas, online education is now possible. Educators or students no longer need to travel to far-flung areas to teach their lessons.

### Constant Connectivity

Constant connectivity ensures that people can respond to emergencies relatively quickly. For example, a wireless device like mobile can ensure you a constant connectivity though you move from place to place or while you travel, whereas a wired landline can't.

## Mobile Communication Protocols

Any device that does not need to remain at one place to carry out its functions is a mobile device. So laptops, smartphones and personal digital assistants are some examples of mobile devices. Due to their portable nature, mobile devices connect to networks wirelessly. Mobile devices typically use radio waves to communicate with other devices and networks.

Mobile communication protocols use multiplexing to send information. Multiplexing is a method to combine multiple digital or analog signals into one signal over the data channel. This ensures optimum utilization of expensive resource and time. At the destination these signals are de-multiplexed to recover individual signals.

These are the types of multiplexing options available to communication channels:

FDM (Frequency Division Multiplexing) – Here each user is assigned a different frequency from the complete spectrum. All the frequencies can then simultaneously travel on the data channel.

TDM (Time Division Multiplexing) – A single radio frequency is divided into multiple slots and each slot is assigned to a different user. So multiple users can be supported simultaneously.

CDMA (Code Division Multiplexing) – Here several users share the same frequency spectrum simultaneously. They are differentiated by assigning unique codes to them. The receiver has the unique key to identify the individual calls.

### GSM

GSM stands for Global System for Mobile communications. GSM is one of the most widely used digital wireless telephony system. It was developed in Europe in 1980s and is now international standard in Europe, Australia, Asia and Africa. Any GSM handset with a SIM (Subscriber Identity Module) card can be used in any country that uses this standard. Every SIM card has a unique identification number. It has memory to store applications and data like phone numbers, processor to carry out its functions and software to send and receive messages.

GSM technology uses TDMA (Time Division Multiple Access) to support up to eight calls simultaneously. It also uses encryption to make the data more secure.

The frequencies used by the international standard is 900 MHz to 1800 MHz However, GSM phones used in the US use 1900 MHz frequency and hence are not compatible with the international system.

### CDMA

CDMA stands for Code Division Multiple Access. It was first used by the British military during World War II. After the war its use spread to civilian areas due to high service quality. As each user gets the entire spectrum all the time, voice quality is very high. Also, it is automatically encrypted and hence provides high security against signal interception and eavesdropping.

### WLL

WLL stands for Wireless in Local Loop. It is a wireless local telephone service that can be provided in homes or offices. The subscribers connect to their local exchange instead of the central exchange wirelessly. Using wireless link eliminates last mile or first mile construction of network connection, thereby reducing cost and set up time. As data is transferred over very short range, it is more secure than wired networks.

WLL system consists of user handsets and a base station. The base station is connected to the central exchange as well as an antenna. The antenna transmits to and receives calls from users through terrestrial microwave links. Each base station can support multiple handsets depending on its capacity.

### GPRS

GPRS stands for General Packet Radio Services. It is a packet based wireless communication technology that charges users based on the volume of data they send rather than the time duration for which they are using the service. This is possible because GPRS sends data over the network in packets and its throughput depends on network traffic. As traffic increases, service quality may go down due to congestion, hence it is logical to charge the users as per data volume transmitted.

GPRS is the mobile communication protocol used by second (2G) and third generation (3G) of mobile telephony. It pledges a speed of 56 kbps to 114 kbps, however the actual speed may vary depending on network load.

# Modulation

A signal can be anything like a sound wave which comes out when you shout. This shout can be heard only up to a certain distance. But for the same wave to travel over a long distance, you'll need a technique which adds strength to this signal, without disturbing the parameters of the original signal.

## What is Signal Modulation?

A message carrying signal has to get transmitted over a distance and for it to establish a reliable communication, it needs to take the help of a high frequency signal which should not affect the original characteristics of the message signal.

The characteristics of the message signal, if changed, the message contained in it also alters. Hence it is a must to take care of the message signal. A high frequency signal can travel up to a longer distance, without getting affected by external disturbances. We take the help of such high frequency signal which is called as a carrier signal to transmit our message signal. Such a process is simply called as Modulation. Modulation is the process of changing the parameters of the carrier signal, in accordance with the instantaneous values of the modulating signal.

## Need for Modulation

The baseband signals are incompatible for direct transmission. For such a signal, to travel longer distances, its strength has to be increased by modulating with a high frequency carrier wave, which doesn't affect the parameters of the modulating signal.

## Advantages of Modulation

The antenna used for transmission, had to be very large, if modulation was not introduced. The range of communication gets limited as the wave cannot travel to a distance without getting distorted.

Following are some of the advantages for implementing modulation in the communication systems.

- Antenna size gets reduced.
- No signal mixing occurs.
- Communication range increases.
- Multiplexing of signals occur.
- Adjustments in the bandwidth is allowed.
- Reception quality improves.

## Signals in the Modulation Process

Following are the three types of signals in the modulation process:

- Message or Modulating Signal: The signal which contains a message to be transmitted, is called as a message signal. It is a baseband signal, which has to undergo the process of modulation, to get transmitted. Hence, it is also called as the modulating signal.
- Carrier Signal: The high frequency signal which has a certain phase, frequency, and amplitude but contains no information, is called a carrier signal. It is an empty signal. It is just used to carry the signal to the receiver after modulation.
- Modulated Signal: The resultant signal after the process of modulation, is called as the modulated signal. This signal is a combination of the modulating signal and the carrier signal.

## Types of Modulation

There are many types of modulations. Depending upon the modulation techniques used, they are classified as shown in the following figure.

The types of modulations are broadly classified into continuous-wave modulation and pulse modulation.

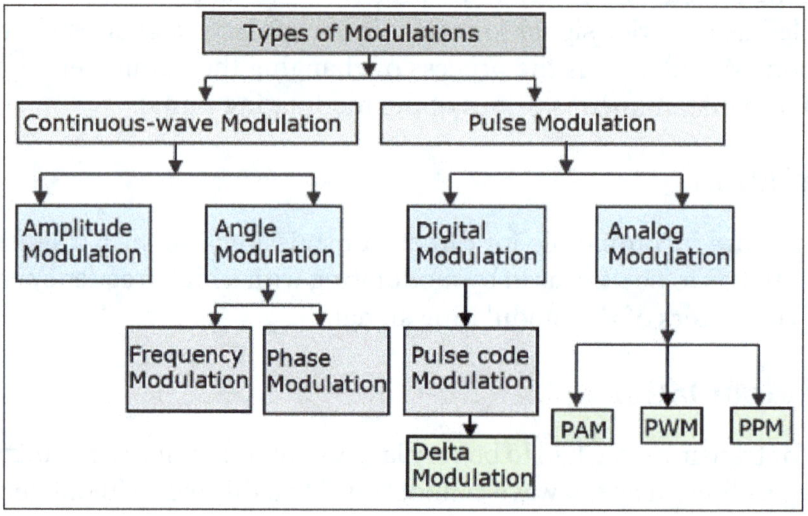


## Continuous-Wave Modulation

In the continuous-wave modulation, a high frequency sine wave is used as a carrier wave. This is further divided into amplitude and angle modulation:

- If the amplitude of the high frequency carrier wave is varied in accordance with the instantaneous amplitude of the modulating signal, then such a technique is called as Amplitude Modulation.
- If the angle of the carrier wave is varied, in accordance with the instantaneous value of the modulating signal, then such a technique is called as Angle Modulation.

The angle modulation is further divided into frequency and phase modulation:

- If the frequency of the carrier wave is varied, in accordance with the instantaneous value of the modulating signal, then such a technique is called as Frequency Modulation.
- If the phase of the high frequency carrier wave is varied in accordance with the instantaneous value of the modulating signal, then such a technique is called as Phase Modulation.

## Pulse Modulation

In Pulse modulation, a periodic sequence of rectangular pulses, is used as a carrier wave. This is further divided into analog and digital modulation.

In analog modulation technique, if the amplitude, duration or position of a pulse is varied in accordance with the instantaneous values of the baseband modulating signal, then such a technique is called as Pulse Amplitude Modulation (PAM) or Pulse Duration/Width Modulation (PDM/PWM), or Pulse Position Modulation (PPM).

In digital modulation, the modulation technique used is Pulse Code Modulation (PCM) where the analog signal is converted into digital form of 1s and 0s. As the resultant is a coded pulse train, this is called as PCM.

## Analog Modulation Methods

Analog modulation refers to the process of transferring an analog baseband (low frequency) signal, like an audio or TV signal over a higher frequency signal such as a radio frequency band. There are two ways to modulate an RF carrier:

## Amplitude Modulation

In analog modulation, the amplitude of the carrier signal is made to follow that of the modulating signal. Several variants of amplitude modulation are used in practice. They are Double Side Band Suppressed Carrier (DSBSC) Modulation, Single Sideband Suppressed Carrier (SSBSC) Modulation and Vestigial Sideband Amplitude Modulation (VSBAM).

## PBDDS Board Implementation

For generation of AM waveforms on SpinCore PulseBlasterDDS Boards, the basic form of amplitude modulation is given by:

AM (t) = Ca*sin(wc*t)*[A+(Ma*sin(wm*t))]

For this formula:

- t = time,
- Ca = amplitude of Carrier waveform (1 Vp-p here),

- wc = angular frequency of the carrier signal in radians/sec,
- wm = angular frequency of the modulating signal in radians/sec,
- Ma = amplitude of the modulating waveform,
- AM(t) = the resulting AM waveform.

A and Ma are set and scaled so that the amplitude of [A+(Ma*Sin(wmt))] does not exceed the value of 1Vp-p for the given modulation index.

The scaling factor and the values are chosen by the formula given below:

Ma = 1/((100/MI) + 1)

Where MI=Modulation Index specified by the user in between 0-100% value. And A = Ma*100/MI

Tm=1/Fm can not be less than (9 * clock time period) for proper results where Fm is the frequency of the message signal in Hz.

The carrier waveform and the message signals can be generated using the PulseBlasterDDS and RadioProcessor's NCO and AWG respectively.

The generation of carrier waveform in C code is as below :

```
void shape_make_carrier (float *dds_data)
{
  int i;

  for (i = 0; i < 1024; i++)
    {
      dds_data[i] = sin (2.0 * pi * ((float) i / 1024.0));
    }
}
```

The message signal in which amplitude is scaled in accordance with modulation index is generated as given below.

```
void shape_make_sin (float *shape_data)
{
  int i;
  float MI, A, Ma;
  printf ("Enter Modulation Index from 0 to 100 percent: ");
  scanf ("%f", &MI);

  Ma = 1/((100/MI) + 1);
  A = Ma*100/MI ;

  for (i = 0; i < 1024; i++)
    {
```

```
        shape_data[i] = A+(Ma*sin (2.0 * pi * ((float) i / 1024.0)));
    }
}
```

Both the above generated signals are loaded in the board by using the SpinAPI function as given below:

```
pb_dds_load (shape_data, DEVICE_SHAPE);
pb_dds_load (dds_data, DEVICE_DDS);
```

The complete C code demonstrating this implementation is available for direct download. The code generates an amplitude modulated wave for any given modulation index, carrier frequency, and message signal frequency. In the code, the message signal is assumed to be a sine wave, but the user can edit the code to make the message signal into any waveform.

Some examples of output obtained with this code is given below. The code only demonstrates a basic version of amplitude modulation. A similar setup can also be used for the DSBSC, SSBSC and VSBAM techniques mentioned above.

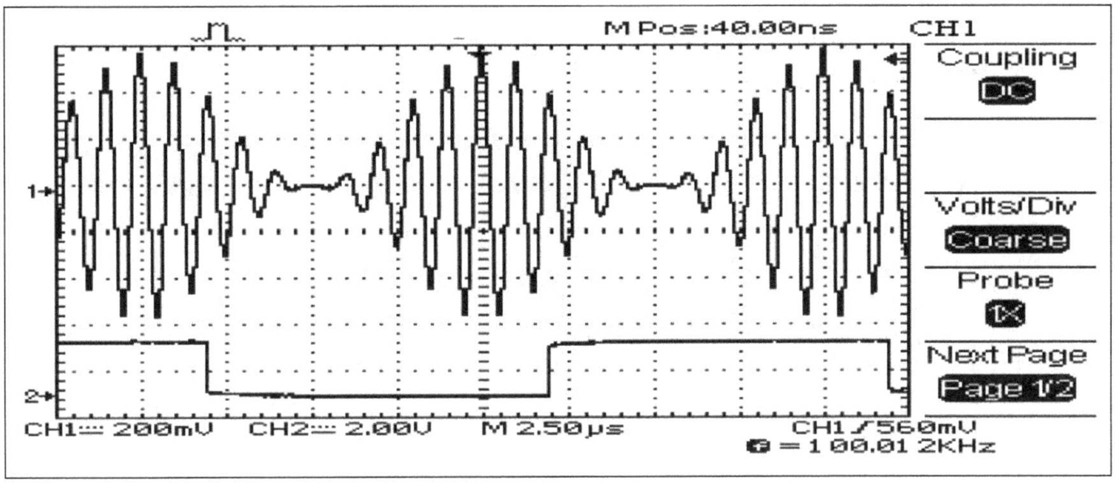


Figure: The output AM waveform generated.

Channel 1 shows the ouput AM wave for carrier frequency Fc =1 MHz and a message signal frequency of Fm = 100 kHz and modulation index = 100%.

Channel 2 shows the TTL output with frequency = Fm/2. Each cycle of the TTL output corresponds to the time taken to execute the two SpinAPI functions given below:

```
pb_inst_radio_shape (0, 0, 0, 0, TX_ENABLE, NO_PHASE_RESET, NO_TRIGGER,
USE_SHAPE, 0, 0x0F, CONTINUE, 0, (1/fm) * us);

pb_inst_radio_shape (0, 0, 0, 0, TX_ENABLE, NO_PHASE_RESET, NO_TRIGGER,
USE_SHAPE, 0, 0x00, BRANCH, start, (1/fm) * us);
```

The execution of these instructions results in the generation of two full cycles of the AM signal.

Also Channel 1 is delayed by (9 * clock cycle) time period.

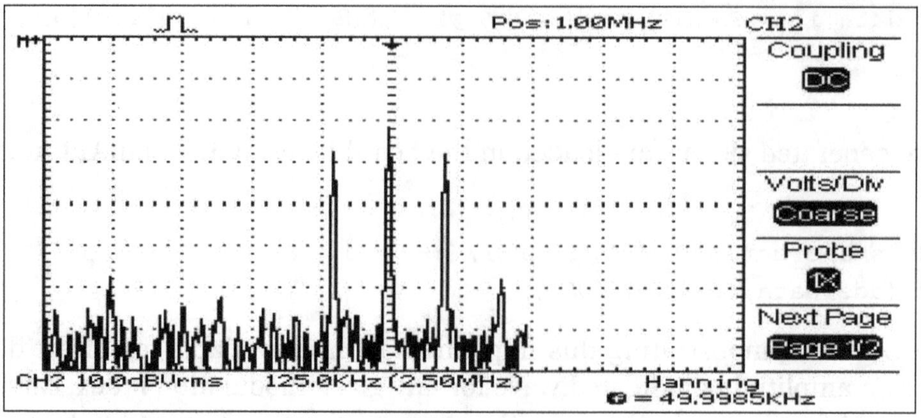


Figure: The frequency spectrum of the output AM waveform for the specifications.

The carrier is centered around at 1 MHz and two sidebands are present at 900 kHz and 1100 kHz. The sidebands are offset from the carrier by the frequency of the modulating signal.

This basic variant of AM has a prominent carrier signal displayed in the frequency spectrum. Some of the other methods of achieving AM mentioned, such as DSBSC and SSBSC above are designed to reduce this carrier component in the frequency spectrum. Methods such as SSBSC and VSBAM are designed to remove or suppress one of the sidebands.

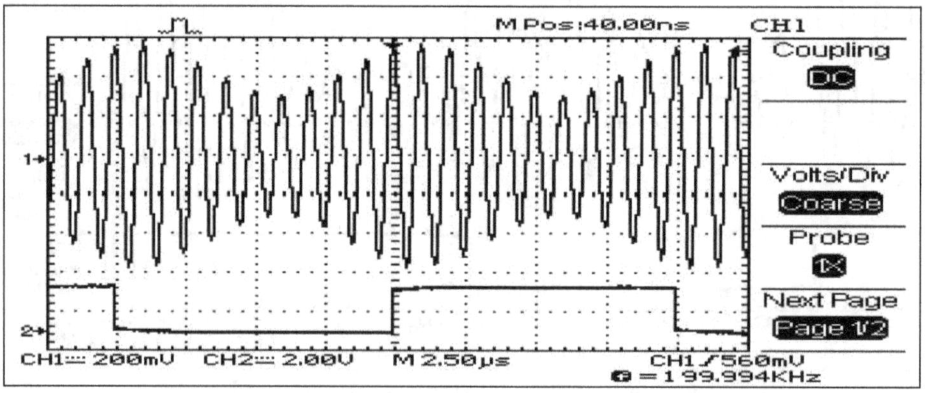


Figure: An AM waveform generated using the example code.

Channel 1 shows the ouput AM wave for a carrier frequency Fc = 1 MHz and message signal frequency Fm = 100 kHz with Modulation Index = 30%.

Channel 2 shows the TTL output with frequency = Fm/2.

## Angle Modulation

In Angle Modulation, the message signal's amplitude is used to control the frequency or phase of the carrier signal. This gives rise to the two methods known as Frequency Modulation and Phase Modulation, respectively.

## Frequency Modulation Implementation on SpinCore PulseBasterDDS Boards

In the SpinCore PulseBasterDDS board, the frequency modulated waveform is generated using the

NCO by controlling the frequency registers in accordance with the instantaneous amplitude of the message signal using the PulseBlaster Timing Core.

The code performs basic frequency modulation using a sinusoidal carrier for any given carrier frequency within the board specifications. The message signal is also assumed to be a sine wave, as in the case of amplitude modulation. This code can be extended to use different types of carrier and message signals.

The equation implemented in the code is:

FM(t)= Ca*sin(wc*t + phi)

For this formula:

- phi = change in the frequency of the carrier with respect to amplitude of modulating waveform.
- wc = angular frequency of the carrier signal in radians/sec (also equal to 2*pi*fc).
- Ca = amplitude of the carrier signal (set to 1 in the code).
- FM(t) = the resulting FM waveform.

The modulating signal chosen is a sine wave as given by:

m(t) = Ma*sin(wm*t)

For this formula:

- m(t) = the modulating or message signal.
- wm = angular frequency of the message signal in radians/sec (also equal to 2*pi*fm).
- Ma = amplitude of the modulating signal (set to 1 in the code).

Also, in this example, the TTL outputs are used to trigger the oscilloscope.

Following the similar method described in amplitude modulation, the carrier and message signal is generated in the C code.

In this implementation, the modulating waveform amplitude (which is in the range of -1V to +1V) is quantized into a number of value as set by the user (with greater numbers offering a better FM implementation). These values are in turn used to compute one of that many possible frequency shifts around the carrier frequency. This is done by choosing a step size such that a +1V amplitude would produce a shift resulting in a instantaneous carrier frequency of fs + (fm/2) and a -1V would result in fs - (fm/2). The quantization size should be set to the number of frequency registers available on the particular SpinCore PulseBlasterDDS or RadioProcessor board being used. If more frequency registers are desired, please contact SpinCore.

The frequency registers can be loaded with the required value by using the following SpinAPI functions:

```
for (i=0;i<N;i++)
```

```
{
        pb_set_freq (fc[i]);

}
```

where fc[i] contains the values of the frequency shift in accordance to the quantized amplitude.

The SpinAPI functions used for frequency modulation is given by:

```
pb_inst_radio(i,0,0,0, TX_ENABLE, NO_PHASE_RESET, NO_TRIGGER, 0xFF,CONTINUE,
0, Ti*ns);
```

where the index "i" represents the ith frequency register.

Some examples of output obtained with this code is given below. Note that the code only demonstrates a basic version of frequency modulation. It can also be used to implement the phase modulation by using the phase registers of the PulseBlasterDDS and RadioProcessor boards instead of the frequency registers.

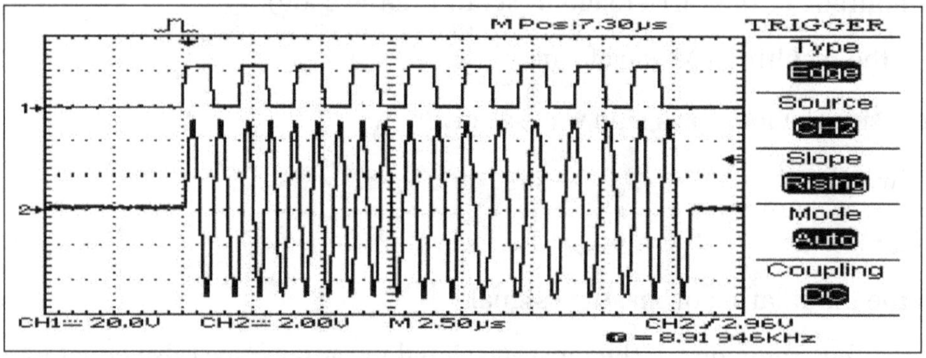


Figure: In this example, for the purpose of visualizing the results on the oscilloscope, the instantaneous frequency of the carrier signal is quantized to only 9 possible values.

The figure shows the output FM waveform on channel 2 with TTL outputs on Channel 1 being used for triggering the oscilloscope for a carrier frequency of Fc = 1 MHz and a message signal frequency of Fm = 500 kHz.

Each pulse in the TTL waveform corresponds to the execution of an instruction that generates a portion of the FM signal corresponding to 1/9th of the period of the message signal.

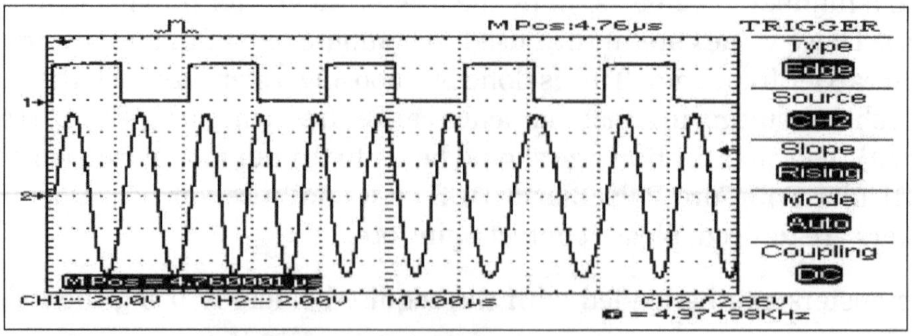


Figure: Is the result obtained with the same parameters that were used in the previous example. Here, the number of frequency values for the carrier signal was quantized to 5 values to give a even better visualization of the frequency modulation.

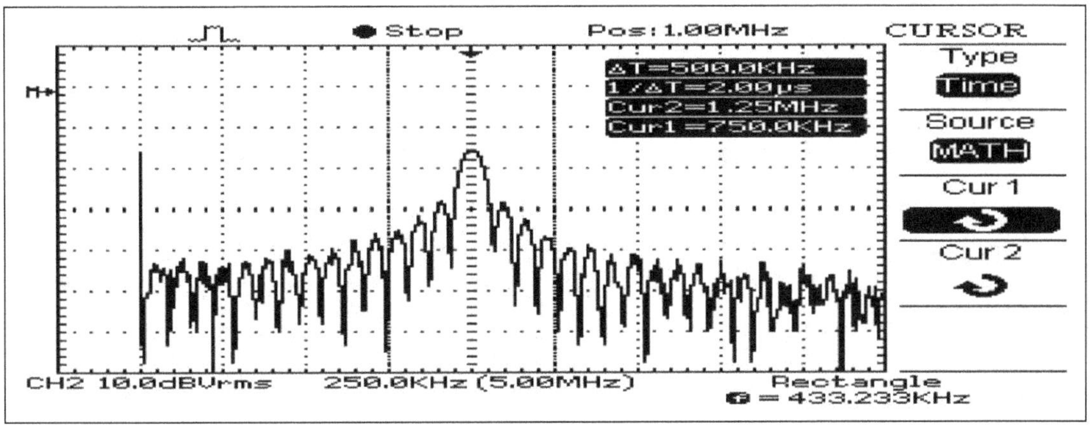


Figure: The FM spectrum on channel 2 of the oscilloscope when 1024 frequency values are used in the carrier signal. The same parameters as the examples given above were used for this example as well.

## Digital Modulation Methods

The term DM stands for digital modulation, and it is a common term for the techniques of modulation. This modulation uses discrete signals for modulating a carrier wave. Indifference, both the amplitude modulation and frequency modulation techniques are analog. Digital modulation removes communication noise as well as provides enhanced strength for the signal intrusion. But, it is not rare to digital modulation schemes for introducing time delay because of the required process. To avoid this, a comfort SST (Secure Stream Technology) audio is designed.

Digital Modulation (DM) gives more capacity of data, high information security, and accessibility of a faster system by enormous quality communication. Therefore, DM techniques have a huge demand due to their capacity for communicating superior amounts of information than AM (analog modulation) techniques.

## Types of Digital Modulation

There are several kinds of digital modulation techniques are available based on the requirement which includes the following:

- ASK or Amplitude shift Key.
- FSK or Frequency shift key.
- PSK or Phase shift key.

## Difference between Analog and Digital Modulation

The difference between Analog Modulation and Digital Modulation include the following:

| Analog Modulation | Digital Modulation |
|---|---|
| An AM signal can signify any value in a range. | A DM signal can only signify with a set of discrete values. |
| In analog modulation (AM), the input must be in the form of analog. | In digital modulation (DM), the input must be the data in the form of digital. |

| | |
|---|---|
| In AM, the value between the max & min is considered to be applicable. | In DM, only two binary numbers are considered applicable such as 1 and 0. |
| Most of the signals that we transmit in nature are analog like voice signal, and it is much simpler to complete analog modulation compare with digital. | But using digital modulation, you require transmitting through an ADC converter before transmission & a DAC at the end of the receiver for recovering the unique signal. The extra phases required to transmit DM (digital modulation) enhances both the price and difficulty of the transmitter as well as receiver. |
| The AM can generate a signal to carry the frequently changing data. | The DM generates a signal whose rate changes at particular time intervals. |
| In AM, it is not easy to disconnect the signal from noise. | In DM, the signal can simply disconnect from noise. |

This modulation offers more capacity of data, digital data services with compatibility; data security is high, good communication quality, and faster system accessibility. These modulation schemes have a better capacity for conveying a huge amount of data than AM schemes (analog modulation).

### Phase- Shift Keying

Phase shift keying, PSK, is widely used these days within a whole raft of radio communications systems. It is particularly well suited to the growing area of data communications. PSK, phase shift keying enables data to be carried on a radio communications signal in a more efficient manner than Frequency Shift Keying, FSK, and some other forms of modulation.

With more forms of communications transferring from analogue formats to digital formats, data communications is growing in importance, and along with it the various forms of modulation that can be used to carry data.

There are several flavours of phase shift keying, PSK that are available for use. Each form has its own advantages and disadvantages, and a choice of the optimum format has to be made for each radio communications system that is designed. To make the right choice it is necessary to have a knowledge and understanding of the way in which PSK works.

### Phase Shift Keying, PSK and Basics

Like any form of shift keying, there are defined states or points that are used for signalling the data bits. The basic form of binary phase shift keying is known as Binary Phase Shift Keying (BPSK) or it is occasionally called Phase Reversal Keying (PRK). A digital signal alternating between +1 and -1 (or 1 and 0) will create phase reversals, i.e. 180 degree phase shifts as the data shifts state.

The problem with phase shift keying is that the receiver cannot know the exact phase of the transmitted signal to determine whether it is in a mark or space condition. This would not be possible even if the transmitter and receiver clocks were accurately linked because the path length would determine the exact phase of the received signal. To overcome this problem PSK systems use a differential method for encoding the data onto the carrier. This is accomplished, for example, by making a change in phase equal to a one, and no phase change equal to a zero. Further improvements can be made upon this basic system and a number of other types of phase shift keying have been developed. One simple improvement can be made by making a change in phase by 90 degrees in one direction for a one, and 90 degrees the other way for a zero. This retains the 180 degree phase reversal between one and zero states, but gives a distinct change for a zero. In a basic system not using this process it may be possible to loose synchronisation if a long series of zeros are sent. This is because the phase will not change state for this occurrence.

There are many variations on the basic idea of phase shift keying. Each one has its own advantages and disadvantages enabling system designers to choose the one most applicable for any given circumstances. Other common forms include QPSK (Quadrature phase shift keying) where four phase states are used, each at 90 degrees to the other, 8-PSK where there are eight states and so forth.

## PSK Constellation Diagrams

It is often convenient to represent a phase shift keyed signal, and sometimes other types of signal using a phasor or constellation diagram. Using this scheme, the phase of the signal is represented by the angle around the circle, and the amplitude by the distance from the origin or centre of the circle. In this way the can be signal resolved into quadrature components representing the sine or I for In-phase component and the cosine for the quadrature component. Most phase shift keyed systems use a constant amplitude and therefore points appear on one circle with a constant amplitude and the changes in state being represented by movement around the circle. For binary shift keying using phase reversals the two points appear at opposite points on the circle. Other forms of phase shift keying may use different points on the circle and there will be more points on the circle.

When plotted using test equipment errors may be seen from the ideal positions on the phase diagram. These errors may appear as the result of inaccuracies in the modulator and transmission and reception equipment, or as noise that enters the system. It can be imagined that if the position of the real measurement when compared to the ideal position becomes too large, then data errors will appear as the receiving demodulator is unable to correctly detect the intended position of the point around the circle.

Using a constellation view of the signal enables quick fault finding in a system. If the problem is related to phase, the constellation will spread around the circle. If the problem is related to magnitude, the constellation will spread off the circle, either towards or away from the origin. These graphical techniques assist in isolating problems much faster than when using other techniques.

QPSK is used for the forward link form the base station to the mobile in the IS-95 cellular system and uses the absolute phase position to represent the symbols. There are four phase decision points, and when transitioning from one state to another, it is possible to pass through the circle's origin, indicating minimum magnitude.

On the reverse link from mobile to base station, O-QPSK is used to prevent transitions through the origin. Consider the components that make up any particular vector on the constellation diagram as X and Y components. Normally, both of these components would transition simultaneously, causing the vector to move through the origin. In O-QPSK, one component is delayed, so the vector will move down first, and then over, thus avoiding moving through the origin, and simplifying the radio's design. A constellation diagram will show the accuracy of the modulation.

### Forms of Phase Shift Keying

Although phase modulation is used for some analogue transmissions, it is far more widely used as a digital form of modulation where it switches between different phases. This is known as phase shift keying, PSK, and there are many flavours of this. It is even possible to combine phase shift keying and amplitude keying in a form of modulation known as quadrature amplitude modulation, QAM.

The list below gives some of the more commonly used forms of phase shift keying, PSK, and related forms of modulation that are used:

- PSK - Phase Shift Keying.
- BPSK - Binary Phase Shift Keying.
- QPSK - Quadrature Phase Shift Keying.
- O-QPSK - Offset Quadrature Phase Shift Keying.
- 8 PSK - 8 Point Phase Shift Keying.
- 16 PSK - 16 Point Phase Shift Keying.

These are just some of the major forms of phase shift keying, PSK, that are widely used in radio communications applications today. Each form of phase shift keying has its own advantages and disadvantages. In general the higher order forms of modulation allow higher data rates to be carried within a given bandwidth. However the downside is that the higher data rates require a better signal to noise ratio before the error rates start to rise and this counteracts any improvements in data rate performance. In view of this balance many radio communications systems are able to dynamically choose the form of modulation depending upon the prevailing conditions and requirements.

### Frequency Shift Keying

The frequency shift keying is the most important digital modulation technique, and it is also known as FSK. A signal has the amplitude, frequency, and phase as properties. Every signal has these three properties. To increase any one of the signal property we can go for the modulation process. Because there are various advantages of the modulation technique. In those some of the advantages are – the antenna size reduced, avoid multiplexing of signals, decrease the SNR, long-range communication can be possible, etc. These are the important advantages of the modulation process. If we modulate the amplitude of the input binary signal according to the carrier signal i.e. called as amplitude shift keying.

It is defined as the changing or improving the frequency characteristics of an input binary signal according to the carrier signal. Amplitude variation is one of the major drawbacks in ASK. So, due to this ask modulation technique used in a few applications only. And its spectrum power efficiency also low. It leads to wastage of power. So to overcome these drawbacks Frequency Shift Keying is preferred. FSK is also known as Binary Frequency Shift Keying (BFSK). The below frequency shift keying theory describes what happening in frequency shift keying modulation.

## Frequency Shift Keying Theory

This frequency shift keying theory shows how the frequency characteristics of a binary signal changed according to the carrier signal. In FSK, the binary information can be transmitted through a carrier signal along with frequency changes. The below diagram shows the frequency shift keying block diagram.

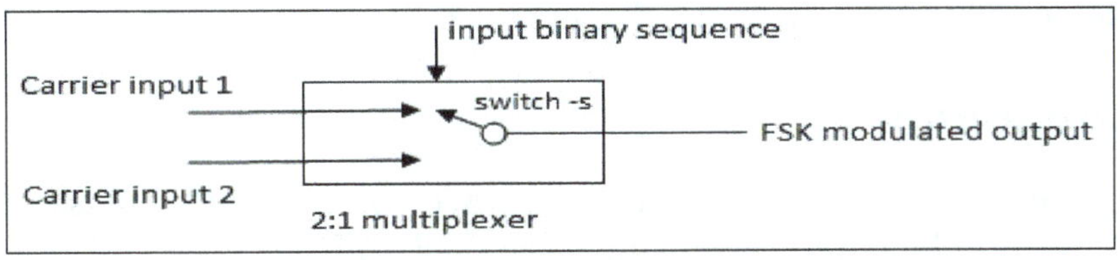


Figure: FSK-block-diagram.

In FSK, two carrier signals are used to produce FSK modulated waveforms. The reason behind this, FSK modulated signals are represented in terms of two different frequencies. The frequencies are called "mark frequency" and "space-frequency". Mark frequency has represented logic 1 and space-frequency has represented the logic 0. There is only one difference between these two carrier signals, i.e. carrier input 1 having more frequency than the carrier input 2.

### Carrier Input 1 = Ac Cos $(2\omega c+\theta)$ t

### Carrier Input 2 = Ac Cos $(2\omega c-\theta)$ t

The switch (s) of the 2:1 multiplexer is having the important role to generate the FSK output. Here the switch is connected to carrier input 1 for all logic 1's of the binary input sequence. And switch (s) is connected to carrier input 2 for all logic 0's of the input binary sequence. So, the resultant FSK modulated waveforms have mark frequencies and space frequencies.

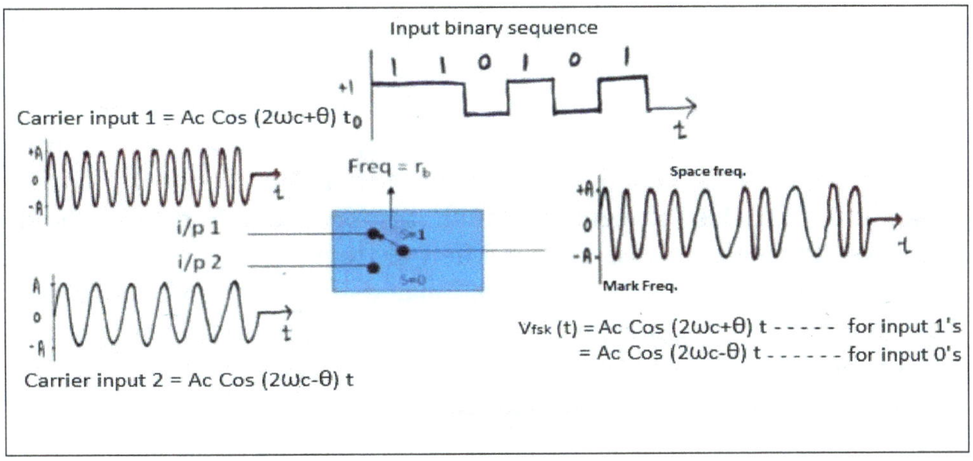


Figure: FSK-modulation-output-waveforms.

Now we will see how the FSK modulated wave can be demodulated at the receiver side. Demodulation is defined as reconstructing the original signal from the modulated signal. This demodulation can be possible in two ways. They are:

- Coherent FSK detection.
- Non-coherent FSK detection.

The only difference between the coherent and non-coherent way of detection is the phase of the carrier signal. If the carrier signal we are using at the transmitter side and receiver side are in the same phase while demodulation process i.e. called a coherent way of detection and it is also known as synchronous detection. If the carrier signals which we are using at transmitter and receiver side are not in the same phase then such modulation process known as Non-coherent detection. Another name for this detection is Asynchronous detection.

## Coherent FSK Detection

In this synchronous FSK detection, the modulated wave got affected by noise while reaching the receiver. So, this noise can be eliminated from using the bandpass filter (BPF). Here at multiplier stage, the noisy FSK modulated signal is multiplied with the carrier signal from the local oscillator device. Then the resultant signal passes from the BPF. Here this bandpass filter is assigned to cut off frequency which is equal to the binary input signal frequency. So the same frequencies can be allowed to the decision device. Here this decision device gives 0 and 1 for space and mark frequencies of the FSK modulated waveforms.

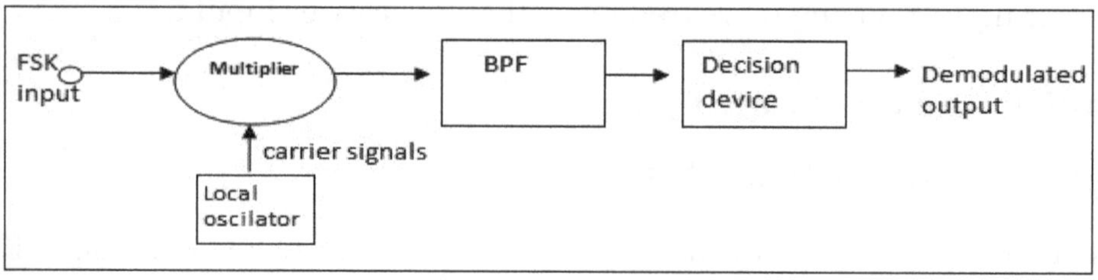


Figure: Coherent-FSK-detection.

## Non-coherent FSK Detection

The modulated FSK signal is forwarded from the bandpass filter 1 and 2 with cut off frequencies equals to space and mark frequencies. So, the unwanted signal components can be eliminated from the BPF. And the modified FSK signals are applied as input to the two envelop detectors. This envelope detector is a circuit having a diode (D). Based upon the input to the envelope detector it delivers the output signal. This envelope detector used in the amplitude demodulation process. Based upon its input it generates the signal and then it is forwarded to the threshold device. This threshold device gives the logic 1 and 0 for the different frequencies. This would be equal to the original binary input sequence. So, the FSK generation and detection can be done in this way. This process can be known for the frequency-shift keying modulation and demodulation experiment also. In this FSK experiment, FSK can be generated by the 555 timer IC and detection can be possible by 565IC which is known as a phase-locked loop (PLL).

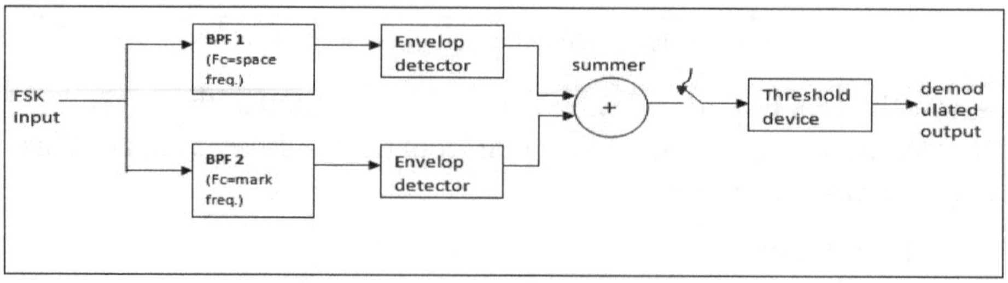


Figure: Non-coherent-FSK-detection.

There are few frequency shift keying advantages and disadvantages are listed below:

## Advantages

- Simple process to construct the circuit.
- Zero amplitude variations.
- Supports a high data rate.
- Low probability of error.
- High SNR (signal to noise ratio).
- More noise immunity than the ASK.
- Error-free reception can be possible with FSK.
- Useful in high-frequency radio transmissions.
- Preferable in high-frequency communications.
- Low-speed digital applications.

## Disadvantages

- It requires more bandwidth than the ASK and PSK(phase shift keying).
- Due to the requirement of large bandwidth, this FSK has limitations to use only in low-speed modems which the bit rate is 1200bits/sec.
- The bit error rate is less in AEGN channel than phase shift keying.

Thus, the frequency shift keying is one of the fine digital modulation techniques to increase the frequency characteristics of the input binary signal. By FSK modulation technique we can achieve error-free communication in a few digital applications. But this FSK has finite data rate and consumes more bandwidth can be overcome by the QAM, which is known as quadrature amplitude modulation. It is the combination of amplitude modulation and phase modulation.

## Preparation

## Generation

As its name suggests, a frequency shift keyed transmitter has its frequency shifted by the message.

Although there could be more than two frequencies involved in an FSK signal, the message will be a binary bit stream, and so only two frequencies will be involved.

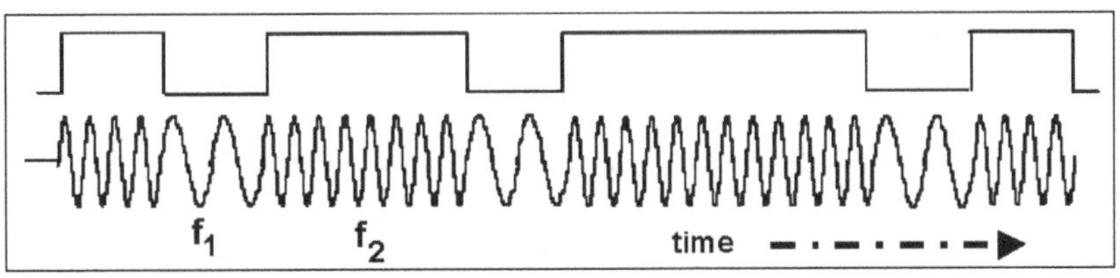


Figure: FSK waveform, derived from a binary message.

The word 'keyed' suggests that the message is of the 'on-off' (mark-space) variety, such as one (historically) generated by a morse key, or more likely in the present context, a binary sequence. The output from such a generator is illustrated in figure below.

Conceptually, and in fact, the transmitter could consist of two oscillators (on frequencies f1 and f2), with only one being connected to the output at any one time. This is shown in block diagram form in Figure below.

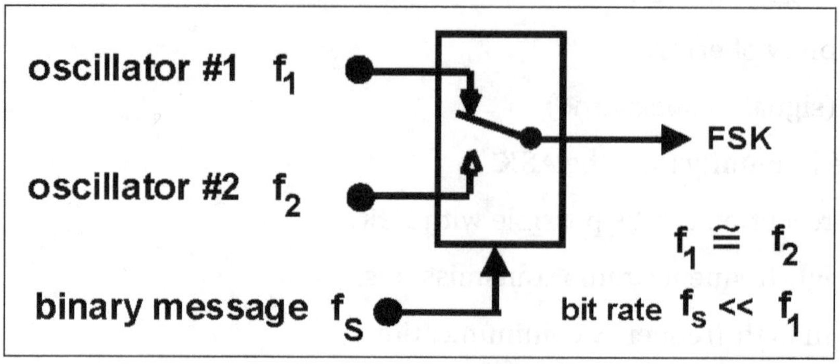


Figure: FSK transmitter.

Unless there are special relationships between the two oscillator frequencies and the bit clock there will be abrupt phase discontinuities of the output waveform during transitions of the message.

## Bandwidth

Practice is for the tones f1 and f2 to bear special inter-relationships, and to be integer multiples of the bit rate. This leads to the possibility of continuous phase, which offers advantages, especially with respect to bandwidth control. Alternatively the frequency of a single oscillator (VCO) can be switched between two values, thus guaranteeing continuous phase - CPFSK. The continuous phase advantage of the VCO is not accompanied by an ability to ensure that f1 and f2 are integer multiples of the bit rate. This would be difficult (impossible?) to implement with a VCO. Being an example of non-linear modulation, calculation of the bandwidth of an FSK signal is a non-trivial exercise.

FSK signals can be generated at baseband, and transmitted over telephone lines (for example). In this case, both f1 and f2 would be audio frequencies. Alternatively, this signal could be translated to a higher frequency. Yet again, it may be generated directly at 'carrier' frequencies.

## Demodulation

There are different methods of demodulating FSK. A natural classification is into synchronous (coherent) or asynchronous (non-coherent). Representative demodulators of these two types are the following:

## Asynchronous

A close look at the waveform of Figure reveals that it is the sum of two amplitude shift keyed (ASK) signals. The receiver of Figure takes advantage of this. The FSK signal has been separated into two parts by bandpass filters (BPF) tuned to the MARK and SPACE frequencies.

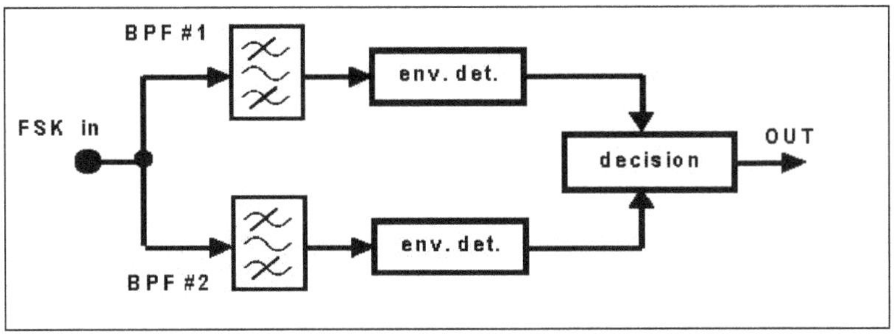


Figure: Demodulation by conversion-to-ASK.

The output from each BPF looks like an amplitude shift keyed (ASK) signal.

These can be demodulated asynchronously, using the envelope. The decision circuit, to which the outputs of the envelope detectors are presented, selects the output which is the most likely one of the two inputs. It also re-shapes the waveform from a bandlimited to a rectangular form.

This is, in effect, a two channel receiver. The bandwidth of each is dependent on the message bit rate. There will be a minimum frequency separation required of the two tones.

## Hint

You are advised to read ahead, before attempting the experiment, to consider the modelling of this demodulator. Unlike most TIMS models, you are not free to choose parameters - particularly frequencies. If they are to be tuned to different frequencies, then one of these frequencies must be 2.083 kHz (defined as the MARK frequency). This is a restriction imposed by the BIT CLOCK REGEN module, of which the BPF are sub-systems. As a result of this, most other frequencies involved are predetermined. Make sure you appreciate why this is so, then decide upon:

- Bit clock rate.
- SPACE frequency.
- Envelope detector LPF characteristics.

## Synchronous

In the block diagram of Figure two local carriers, on each of the two frequencies of the binary FSK signal, are used in two synchronous demodulators. A decision circuit examines the two outputs, and decides which is the most likely.

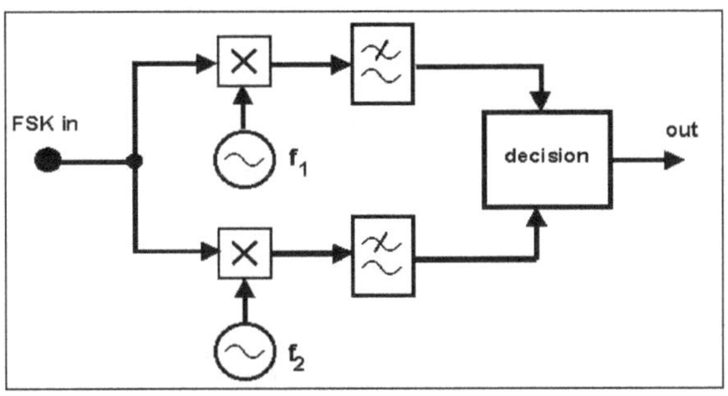


Figure: Synchronous demodulation.

This is, in effect, a two channel receiver. The bandwidth of each is dependent on the message bit rate. There will be a minimum frequency separation required of the two tones. This demodulator is more complex than most asynchronous demodulators.

### Phase Locked Loop

A phase locked loop is a well-known method of demodulating an FM signal. It is thus capable of demodulating an FSK signal. It is shown, in block diagram form, in Figure below.

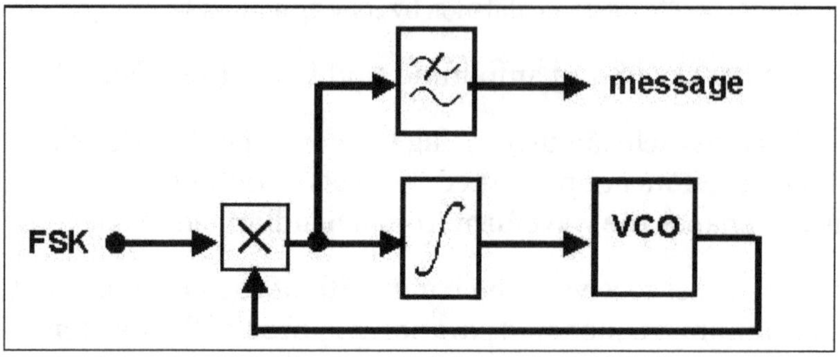


Figure: Phase locked loop demodulator.

The control signal, which forces the lock, is a bandlimited copy of the message sequence. Depending upon the bandwidth of the loop integrator, a separate LPF will probably be required (as shown) to recover the message.

### Post-Demodulation Processing

The output of a demodulator will typically be a bandlimited version of the original binary sequence. Some sort of decision device is then required to regenerate the original binary sequence. This is shown in the block diagrams above, but has not been implemented in the TIMS models to follow.

### Comments

One might imply, from all of the above, that the generation and demodulation of an FSK signal is relatively trivial, and that there is not a lot more to know about its properties. Such is not the case.

Extensive research has been carried out into the properties of an FSK signal. This includes the determination of the optimum relationship between the frequencies of the two tones and the data rate. You should refer to your text book for more information.

This experiment is not typical. There are no specific tasks to be completed. Instead you are invited to investigate any or all of the models below in your own way. Various methods of FSK generation are possible with TIMS, and some suggestions follow.

### Generation

### Scheme

A VCO module is ideally suited for the generation of a continuous phase FSK signal, as shown in figure.

In FSK mode the VCO is keyed by the message TTL sequence. Internal circuitry results in a TTL HI switching the VCO to frequency f1, while a TTL LO switches it to frequency f2. These two frequencies may be in the audio range (front panel toggle switch LO), or in the 100 kHz range (front panel toggle switch HI).

The frequencies f1 and f2 are set by the on-board variable resistors RV8 and RV7 respectively, while a continuous TTL HI or a TTL LO is connected to the DATA input socket. In FSK mode neither of the front panel rotary controls of the VCO is in operation.

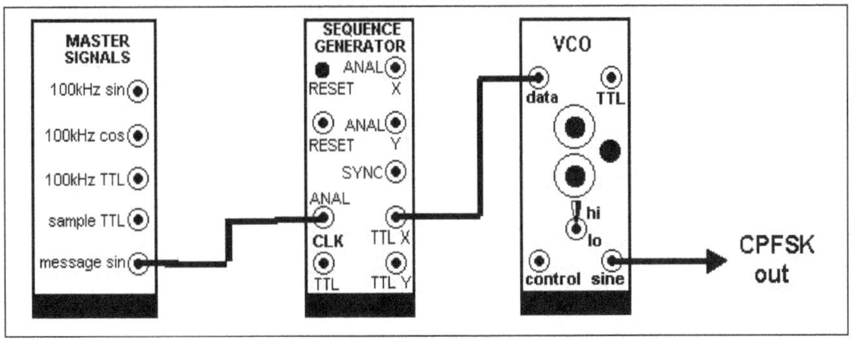


Figure: CPFSK.

## Scheme

Figure shows a model of the arrangement of Figure. It switches either one of two tones to the output, in response to the message sequence.

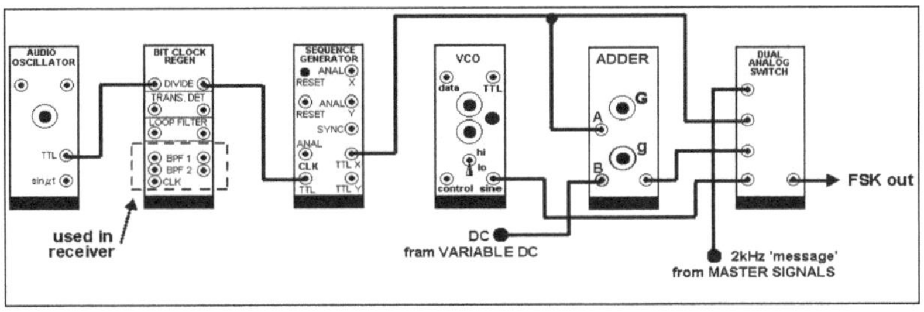


Figure: A model of the arrangement of figure.

The binary sequence is shown clocked by a divided-by-8 version of the output of an AUDIO OSCILLATOR. This oscillator cannot itself be tuned to this relatively low (for TIMS) frequency. The DIVIDE-BY-8 sub-system is in the BIT CLOCK REGEN module (set the on-board switch SW2 with both toggles DOWN).

The signals at f1 and f2 are provided by the 2.083 kHz MESSAGE from the MASTER SIGNALS module, and a VCO, respectively 1. The DUAL AUDIO SWITCH module is used to switch between them.

- One of the two ANALOG SWITCHES is driven directly by the TTL binary message sequence.
- The other ANALOG SWITCH is driven by the same TTL sequence, reversed in polarity, and then DC shifted by +5 volts. The reversal and DC shift is performed by the ADDER, with a maximum -ve output from the VARIABLE DC module. Although 5 volt signals exceed the TIMS ANALOG REFERENCE LEVEL the ADDER design is such that it will not be overloaded.

Unless there is already an FSK signal available at TRUNKS, this transmitter is to be used in conjunction with an asynchronous demodulator, of the type illustrated in Figure, and modelled in Figure - so don`t strip it down unnecessarily.

## Demodulation

In the receivers described below it is assumed there is no bandlimiting (or noise) introduced by a channel. In the case of poor signal-to-noise ratio the MARK and SPACE signals would need to be compared in a decision circuit and the most likely one presented to the output.

## Signals for Demodulation

The demodulators to be examined will require FSK signals as inputs. These may exist at TRUNKS - check. If not, then you will need to generate your own.

For a suitable FSK test signal you could use the model of Figure. The MARK signal is pre-set to 2.083 kHz; initially set the SPACE to about 3 kHz.

## Asynchronous Receiver

An example of this is the demodulator of Figure, shown modelled in figure. The demodulator requires two bandpass (BPF) filters, tuned to the MARK and SPACE frequencies. Suitable filters exist as sub-systems in the BIT CLOCK REGEN module.

To prepare the filters it is necessary to set the on-board switch SW1. Put the left hand toggle UP, and right hand toggle DOWN. This tunes BPF1 to 2.083 kHz, and BPF2 anywhere in the range 1 < fo < 5 kHz, depending on the VCO (the filter centre frequency will be 1/50 of the VCO frequency).

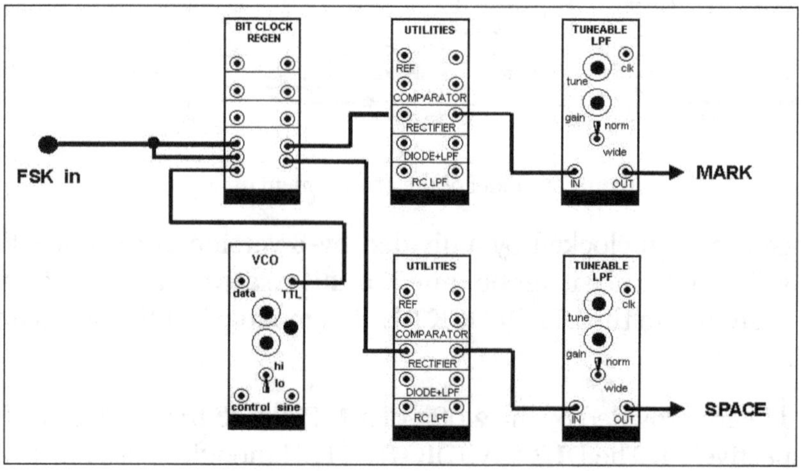


Figure: A model of the receiver of figure.

If you do not have extra UTILITIES and TUNEABLE LPF modules, then complete just one arm of the demodulator.

Alignment requires the BPFs to be tuned to the MARK and SPACE frequencies. The first is already done (2.083 kHz is already pre-set with SW1); the other is set with the VCO (already pre-set with SW2).

Note that the specified bit rate is, by TIMS standards, rather low. The average oscilloscope display can be a little flickery. Use a short sequence, and the SYNC signal from the SEQUENCE GENERATOR to ext. trig.

- What would happen if the bit rate was speeded up ?
- What would happen if the frequency of the SPACE signal, at the transmitter, was moved towards 2.083 kHz ? Of course, the receiver BPF2 would need to be retuned.
- Test you preparatory work and show how close the MARK / SPACE frequencies can approach before performance is degraded - explain why this is so.
- Predict what will happen if the bit rate is increased. If you have supplied your own FSK signal then you should test your prediction.

## Synchronous Receiver

synchronous receiver A synchronous receiver of Figure requires two local carriers, locked to the MARK and SPACE frequencies. Such a receiver would require two VCO and associated modules, and is probably too ambitious to attempt as part of this experiment.

## Pll - Phase Locked Loop

A phase locked loop is shown in block diagram form in figure, and modelled in figure.

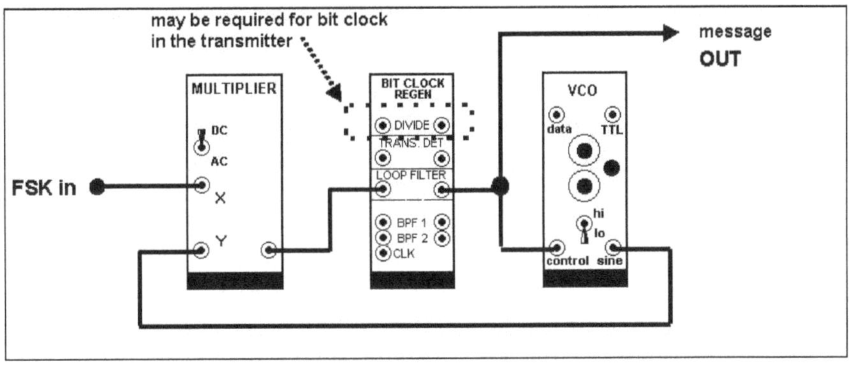


Figure: PLL demodulator - the model of figure.

For the present experiment the integrator is modelled with the LOOP FILTER in the BIT CLOCK REGEN module. This module contains four independent sub-systems. The DIVIDE-BY-8 sub-system may already be in use at the transmitter.

If you are fussy about the appearance of the demodulated output it can be further filtered; say with the LPF in the HEADPHONE AMPLIFIER.

## Quadrature Amplitude Modulation

Quadrature Amplitude Modulation, QAM utilises both amplitude and phase components to provide a form of modulation that is able to provide high levels of spectrum usage efficiency.

QAM, quadrature amplitude modulation has been used for some analogue transmissions including

AM stereo transmissions, but it is for data applications where it has come into its own. It is able to provide a highly effective form of modulation for data and as such it is used in everything from cellular phones to Wi-Fi and almost every other form of high speed data communications system.

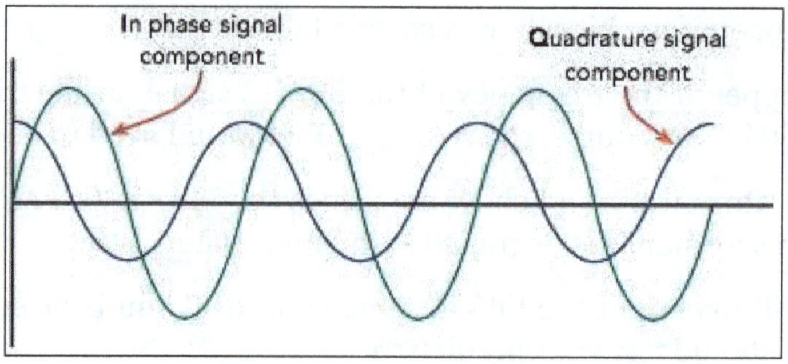


Figure: Quadrature amplitude modulation concept.

Quadrature Amplitude Modulation, QAM is a signal in which two carriers shifted in phase by 90 degrees (i.e. sine and cosine) are modulated and combined. As a result of their 90° phase difference they are in quadrature and this gives rise to the name. Often one signal is called the In-phase or "I" signal, and the other is the quadrature or "Q" signal.

The resultant overall signal consisting of the combination of both I and Q carriers contains of both amplitude and phase variations. In view of the fact that both amplitude and phase variations are present it may also be considered as a mixture of amplitude and phase modulation.

A motivation for the use of quadrature amplitude modulation comes from the fact that a straight amplitude modulated signal, i.e. double sideband even with a suppressed carrier occupies twice the bandwidth of the modulating signal. This is very wasteful of the available frequency spectrum. QAM restores the balance by placing two independent double sideband suppressed carrier signals in the same spectrum as one ordinary double sideband supressed carrier signal.

### Analogue and Digital QAM

Quadrature amplitude modulation, QAM may exist in what may be termed either analogue or digital formats. The analogue versions of QAM are typically used to allow multiple analogue signals to be carried on a single carrier. For example it is used in PAL and NTSC television systems, where the different channels provided by QAM enable it to carry the components of chroma or colour information. In radio applications a system known as C-QUAM is used for AM stereo radio. Here the different channels enable the two channels required for stereo to be carried on the single carrier.

Digital formats of QAM are often referred to as "Quantised QAM" and they are being increasingly used for data communications often within radio communications systems. Radio communications systems ranging from cellular technology as in the case of LTE through wireless systems including WiMAX, and Wi-Fi 802.11 use a variety of forms of QAM, and the use of QAM will only increase within the field of radio communications.

### Digital/Quantised QAM Basics

Quadrature amplitude modulation, QAM, when used for digital transmission for radio communications applications is able to carry higher data rates than ordinary amplitude modulated schemes and phase modulated schemes.

Basic signals exhibit only two positions which allow the transfer of either a 0 or 1. Using QAM there are many different points that can be used, each having defined values of phase and amplitude. This is known as a constellation diagram. The different positions are assigned different values, and in this way a single signal is able to transfer data at a much higher rate.

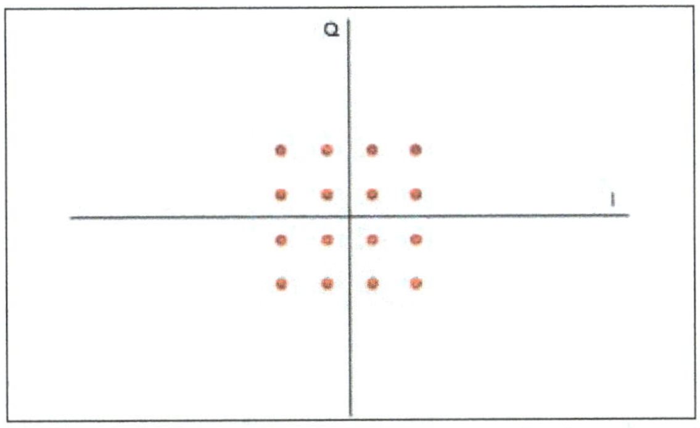


Figure: Constellation diagram for a 16QAM signal showing the location of the different points.

The constellation points are typically arranged in a square grid with equal horizontal and vertical spacing. Although data is binary the most common forms of QAM, although not all, are where there constellation can form a square with the number of points equal to a power of 2 i.e. 4, 16, 64...., i.e. 16QAM, 64QAM, etc.

By using higher order modulation formats, i.e. more points on the constellation, it is possible to transmit more bits per symbol. However the points are closer together and they are therefore more susceptible to noise and data errors.

The advantage of moving to the higher order formats is that there are more points within the constellation and therefore it is possible to transmit more bits per symbol. The downside is that the constellation points are closer together and therefore the link is more susceptible to noise. As a result, higher order versions of QAM are only used when there is a sufficiently high signal to noise ratio.

To provide an example of how QAM operates, the constellation diagram below shows the values associated with the different states for a 16QAM signal. From this it can be seen that a continuous bit stream may be grouped into fours and represented as a sequence.

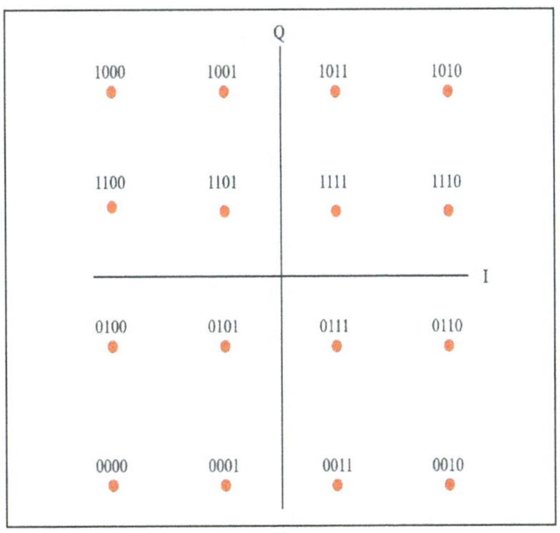


Figure: Bit sequence mapping for a 16QAM signal.

Normally the lowest order QAM encountered is 16QAM. The reason for this being the lowest order normally encountered is that 2QAM is the same as binary phase-shift keying, BPSK, and 4QAM is the same as quadrature phase-shift keying, QPSK.

Additionally 8QAM is not widely used. This is because error-rate performance of 8QAM is almost the same as that of 16QAM - it is only about 0.5 dB better and the data rate is only three-quarters that of 16QAM. This arises from the rectangular, rather than square shape of the constellation.

## QAM Advantages and Disadvantages

Although QAM appears to increase the efficiency of transmission for radio communications systems by utilising both amplitude and phase variations, it has a number of drawbacks. The first is that it is more susceptible to noise because the states are closer together so that a lower level of noise is needed to move the signal to a different decision point. Receivers for use with phase or frequency modulation are both able to use limiting amplifiers that are able to remove any amplitude noise and thereby improve the noise reliance. This is not the case with QAM.

The second limitation is also associated with the amplitude component of the signal. When a phase or frequency modulated signal is amplified in a radio transmitter, there is no need to use linear amplifiers, whereas when using QAM that contains an amplitude component, linearity must be maintained. Unfortunately linear amplifiers are less efficient and consume more power, and this makes them less attractive for mobile applications.

## QAM vs PSK and other Modes

When deciding on a form of modulation it is worth comparing AM vs PSK and other modes looking at what they each have to offer.

As there are advantages and disadvantages of using QAM it is necessary to compare QAM with other modes before making a decision about the optimum mode. Some radio communications systems dynamically change the modulation scheme dependent upon the link conditions and requirements - signal level, noise, data rate required, etc. The table below compares various forms of modulation.

Table: Summary of Types of Modulation with Data Capacities.

| MODULATION | BITS PER SYMBOL | -- ERROR MARGIN -- | | COMPLEXITY |
|---|---|---|---|---|
| OOK | 1 | 1/2 | 0.5 | Low |
| BPSK | 1 | 1 | 1 | Medium |
| QPSK | 2 | $1/\sqrt{2}$ | 0.71 | Medium |
| 16 QAM | 4 | $\sqrt{2}/6$ | 0.23 | High |
| 64QAM | 6 | $\sqrt{2}/14$ | 0.1 | High |

Typically it is found that if data rates above those that can be achieved using 8-PSK are required, it is more usual to use quadrature amplitude modulation. This is because it has a greater distance between adjacent points in the I - Q plane and this improves its noise immunity. As a result it can achieve the same data rate at a lower signal level.

However the points no longer the same amplitude. This means that the demodulator must detect both phase and amplitude. Also the fact that the amplitude varies means that a linear amplifier si required to amplify the signal.

## Fractal Modulation

Fractal modulation is obtained by forming a power weighted superposition of scaled and modulated versions of the signal. The resulting signal is self-similar with fractal characteristics. The wavelet transform can be used as an efficient tool in order to generate a subset of fractal modulated signals that are power homogeneous. Any signal used as a seed for fractal modulation is transformed into a multiscale sound by means of a tree-structured multirate filter bank. Moreover, by superimposing a structured modulation scheme one can generate pseudo-periodic sounds whose partials have fractal behavior.

Introduced in the context of communication, fractal modulation allows for redundant transmission of the same signal at different scales, useful for robust detection in a fading channel scenario or for information hiding, since the characteristics of the fractal modulated signal have a tendency to be more noise-like than the original signal.

The fractal modulation scheme is somehow inspired by the construction of the self-similar function known as the Weierstrass cosine:

$$w(t) = \sum_{n=0}^{+\infty} \gamma^n \cos 2\pi a^n t$$

in which infinite scaled versions of the same cosine function are weighted by a power of γ and added together, where $0 < \gamma < 1 < a$ and $a\gamma \geq 1$. The result is a nowhere differentiable sound whose partials are further and further spaced away, with their frequencies exponentially growing. The Weierstrass cosine is approximately self-similar in the sense that:

$$w(at) = \frac{1}{\gamma}(w(t) - \cos 2\pi t)$$

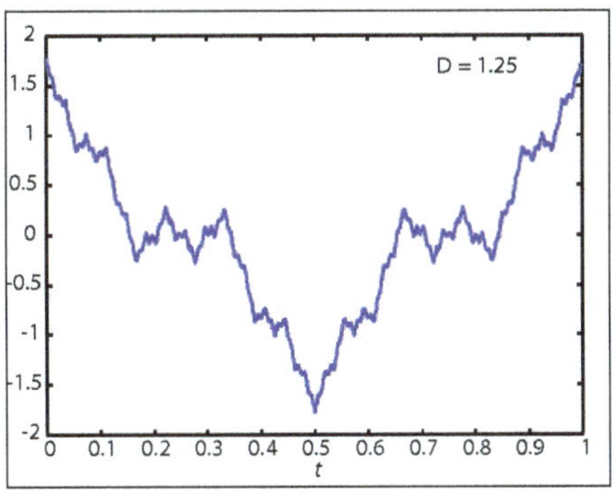


Figure: Weierstrass cosine function with box counting dimension D = 1.25.

so that, time-scaling the function by a factor a obtains the same function w(t), except for a smooth cos $2\pi t$ additional term and an amplitude scaling factor γ. The graph of the Weierstrass cosine, shown in Figure, is a fractal with box counting dimension $D = 2 + \frac{\log \gamma}{\log a}$ log γ log a. Notice that γ controls the fractal dimension: D approaches 2 when γ approaches 1 and decreases with γ.

Similar to the construction of the Weierstrass cosine, fractal modulation employs scaled versions of a signal in order to build a self-similar signal. Moreover, the scaled versions of the signal are modulated to proper band, which is implicit in scaling the cosine functions. The original signal can be considered as a "seed" for the generation of the fractal.

At least in a formal sense, one can construct self-similar signals by adding together a countable number of scaled and modulated versions of the seed. We assume that a > 1 and, usually, 1 < a ≤ 2. Suppose that our seed signal x(t) is bandlimited to [−(a − 1)π, +(a − 1)π], i.e., that its Fourier transform X(ω) is zero for ω outside this interval. We will use the notation $x(t) \in BL_{[A,B]}$ to specify bandlimited signals in the interval [A, B].

By the Fourier scaling theorem, the scaled signal,

$$x_n(t) = \frac{1}{a^n} x\left(\frac{t}{a^n}\right)$$

has Fourier transform,

$$X_n(\omega) = X(a^n \omega)$$

and, therefore,

$$x_n(t) \in BL_{\left[-\frac{(a-1)\pi}{a^n}, +\frac{(a-1)\pi}{a^n}\right]}$$

The peculiar modulation scheme we are going to employ is shown. The scaled signal is split into two subbands, one covering negative frequencies and the other one covering positive frequencies:

$$X_{-}(\omega) = X(\omega)\chi_{]-(a-1)\pi,0]}(\omega)$$

$$X_{+}(\omega) = X(\omega)\chi_{]0,+(a-1)\pi]}(\omega)$$

where χ]A,B](ω) is the characteristic function of the semiclosed interval ]A, B], i.e.

$$\chi_{]A,B]}(\omega) = \begin{cases} 1 & A < \omega \le B \\ 0 & \text{otherwise} \end{cases}$$

The negative frequency band X−(ω) is modulated to positive frequencies while the positive frequency band X+(ω) is modulated to negative frequencies. In this way one obtains, for each scaled and modulated version of the signal, a real signal whose effective bandwidth is the same as the baseband signal. The type of modulation is somehow arbitrary; the reason for this given choice will become apparent.

The choice of modulating frequencies:

$$\omega_n^{\mp} = \pm \frac{\pi}{a^{n-1}}$$

guarantees that, ideally, the scaled and modulated signals do not overlap in the frequency domain. In fact, the scaled and modulated signals have disjoint Fourier transforms for any n:

$$X - (a^n\omega - a\pi) = X\,(a^n\omega - a\pi)\chi_{\left]+\frac{\pi}{a^n},+\frac{\pi}{a^{n-1}}\right]}(\omega)$$

$$X + (a^n\omega + a\pi) = X\,(a^n\omega + a\pi)\chi_{\left]-\frac{\pi}{a^n-1},+\frac{\pi}{a^n}\right]}(\omega)$$

In the frequency domain, the self-similar signal s(t) is obtained as a superposition of the signals in. For ω > 0 we can define the Fourier transform of the self-similar signal as follows:

$$s(\omega) = \sum_{n=-\infty}^{+\infty} \beta^n X\,(a^n\omega - a\pi)\chi_{\left]\frac{\pi}{a^n},\frac{\pi}{a^{n-1}}\right]}(\omega)$$

where β > 1 is an arbitrary parameter. Actually β can be expressed in terms of a as follows:

$$\beta = a^{\gamma}\,;\gamma = \log_a \beta$$

For ω < 0 the Fourier transform of the self-similar signal can be obtained from the relationship:

$$S(-\omega) = S^*(\omega),$$

where the symbol ? denotes complex conjugation. Notice that since the terms of the sum in do not overlap, for any given frequency there is only one nonzero term, hence the sum always converges, with a possible singularity at zero frequency if the original signal has a DC component.

The signal s(t) obtained by taking the inverse Fourier transform of is indeed self-similar with respect to scaling by powers of a. In fact, taking into account and substituting ω aK for ω in, by performing a simple index change in the summation one obtains:

$$S\left(\frac{\omega}{a^K}\right) = a^{K\gamma}\,S\,(\omega).$$

for any K ∈ Z. Therefore, in the time domain we have:

$$s\left(a^K t\right) = a^{K(\gamma-1)} s\,(t),$$

which is the required self-similarity property.

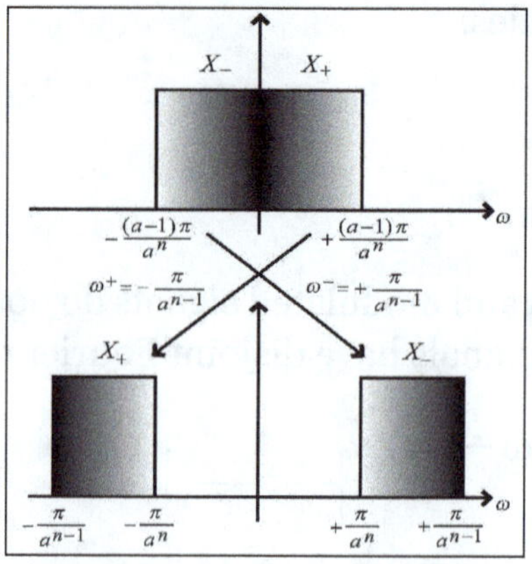


Figure: Modulation scheme employed for the scaled signals.

## Wavelet Based Fractal Modulation

The frequency band subdivision of the scaling-modulation scheme to form self-similar signals illustrated is reminiscent of the structure of wavelet series. Indeed a wavelet based method for fractal modulation was introduced by Wornell and Oppenheim. The simplest form of wavelet series is the expansion:

$$s(t) = \sum_{n,m \in Z} d_n(m)\psi_{n,m}(t)$$

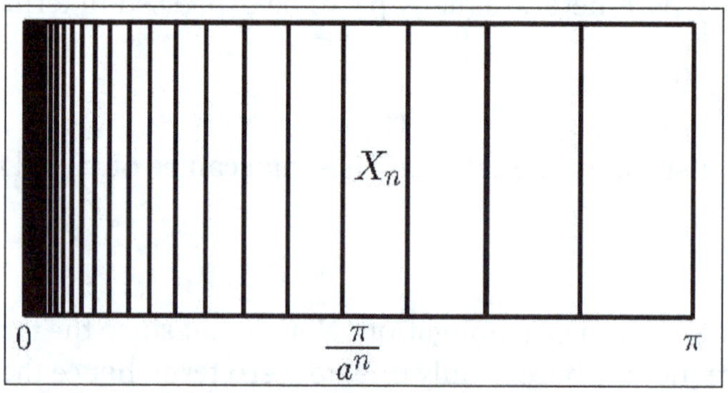


Figure: Frequency band subdivision of the scaled and modulated versions of the signal.

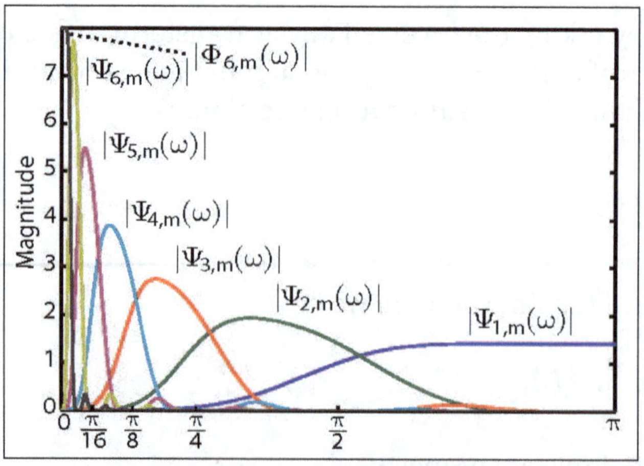


Figure: Magnitude Fourier transform of dyadic wavelets.

of a signal s(t) in terms of a set of functions obtained by timeshifting and scaling a unique function, e.g., ψo,o(t):

$$\psi_{n,n}(t) = \frac{1}{\sqrt{2^n}} \psi_{0,0}\left(\frac{t}{2^n} - m\right); n, m \in \mathbf{Z},$$

where the scaling factor is a = 2. As shown, the dyadic wavelets have bandwidth one octave, with bandwidth decreasing with the scale index n.

Theoretically the wavelets form the tessellation of the timefrequency plane shown in Figure, with higher frequency resolution at large scales and lower frequency resolution at small scales. In practice the cells overlap since the time-frequency uncertainty of the wavelets is larger than the time-frequency sample spacing.

For the orthogonal wavelet bases, the expansion coefficients are obtained by computing the scalar product of the signal with the basis functions ψn,m(t):

$$d_n(m) = \int_{-\infty}^{+\infty} s(t)\psi_{n,m}^*(t)dt$$

Notice that, by performing the variable change t = $2^n$ τ in the integral in and observing, one obtains:

$$d_n(m) = \sqrt{2^n} \int_{-\infty}^{+\infty} s(2^n \tau)\psi_{0,m}^*(\tau)d\tau$$

Suppose that the signal s (t) is dyadically self-similar, i.e., that is satisfied with a = 2, then:

$$d_n(m) = 2^{n\left(\gamma - \frac{1}{2}\right)} d_0(m),$$

i.e., the expansion coefficients of a dyadic self-similar signal on a dyadic wavelet basis are obtained by scaling the amplitude of one and the same coefficient sequence do(m). Vice versa, if the amplitude scaled versions:

$$2^{n\left(\gamma - \frac{1}{2}\right)} x(m)$$

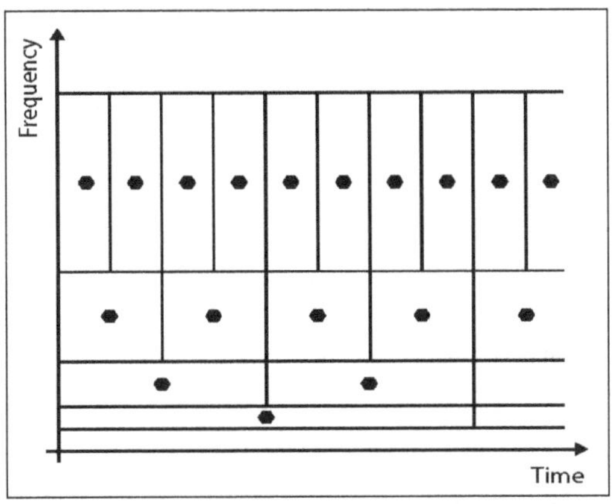


Figure: Time-Frequency cells corresponding to dyadic wavelets.

of a unique discrete-time "seed" signal x(m) are employed as wavelet coefficients, the self-similar signal:

$$s(t)=\sum_{m\in\mathbf{Z}}x(m)\sum_{n\in\mathbf{Z}}2^{n(\gamma-1)}\psi_{0,0}\left(\frac{t}{2^n}-m\right)$$

is obtained. In fact, one can immediately verify that s(t) defined by satisfies with a = 2. Indeed, the functions:

$$\xi_m(t)=\sum_{n\in\mathbf{Z}}2^{n(\gamma-1)}\psi_{0,0}\left(\frac{t}{2^n}-m\right)=\sum_{n\in\mathbf{Z}}2^{n\left(\gamma-\frac{1}{2}\right)}\psi_{n,m}(t)$$

in the linear combination are self-similar.

Intuitively, the wavelet way to fractal modulation is very efficient since, rather than producing scaled and modulated versions of a signal, one pre-scales the known basis functions, an operation that is implicit in the organization of wavelet bases. Moreover, the wavelet series synthesis of a signal can be achieved by means of an iterated two-channel filter bank, where, in each stage, the inputs are up-sampled by a factor 2 and a lowpass filter H(z) and a highpass filter G(z) forming a quadrature mirror filter pair are applied. The diagram in figure serves as a fast generator of self-similar signals.

The computational complexity of the wavelet based fractal modulation scheme depends on the algorithm employed to compute the wavelet transform. However, as we will see, the overall cost is linear with the number of generated samples.

One method to compute fractal modulation is to pre-compute the wavelets at each scale and multiply each of them by the samples of the seed signal. The output is formed by properly timeshifting and adding the pre-multiplied wavelet components. We will refer to this algorithm as the overlap-add method. In this case the complexity can be estimated at N operations per output sample, where N is the number of scales. In fact, the length of the scale n wavelet can be roughly estimated at $2^{n-1}$ L samples obtained as a cascade of upsampling and filtering operators where the order of the filters L–1 is an odd integer for orthogonal wavelets.

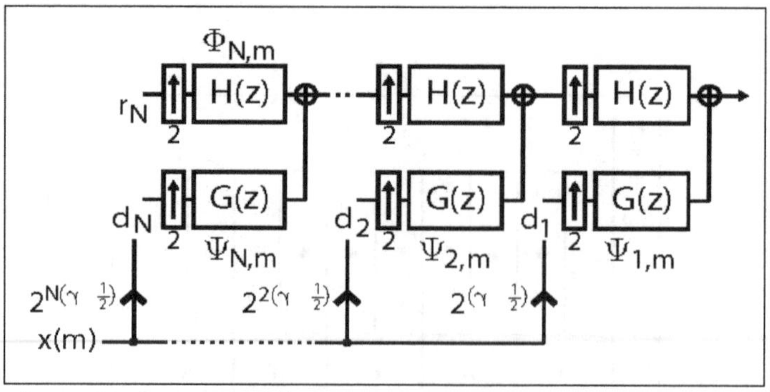


Figure: Wavelet scheme for fractal modulation: an iterated quadrature mirror synthesis filter bank is fed by the same sequence x(m) at any branch. The residue coefficients rN (m) are arbitrary.

Thus, in order to generate an output signal of $2^{N-1}$ L samples one needs to multiply a single largest scale wavelet by one sample of the seed sequence; the two next to largest scale wavelets are each multiplied by one samples of the seed sequence and so forth. In general, computation at scale n requires multiplication of $2^{N-n}$ coefficients by the length $2^{n-1}L$ wavelets. This requires an order of $2^{N-1}L$ operations independently of scale. Therefore the total cost is $2^{N-1}LN$ operations in order to produce $2^{N-1}L$ output samples, which requires a rate of N operations per output sample.

A computational cost of the same order of magnitude is obtained when the scheme in Figure is directly implemented. In this case one can show that the required rate of operations is proportional to the filter length L and does not depend on the number of scales N. It must be pointed out that, for sufficiently large filter orders, FFT based algorithms for the implementation of the filters can be employed to further reduce the overall computational cost.

Due to the finite number of scales obtained by a finite depth filter bank, the synthesized signal is only approximately self-similar. The residue coefficients $r_N(m)$, also known as the coefficients of the scaling function, are arbitrary. If a sufficient number of scales, corresponding to a sufficient number of stages of the filter bank, are employed then the scaling residue represents a very low frequency signal and it can be disregarded in the synthesis of audio signals. However, we will propose the use of self-similar signals also as low-frequency controlling functions of audio effects or as modulators of partials, in which case the scaling residue can be associated either to an amplitude scaled version of the seed signal or to an envelope function.

An important property of the wavelet based self-similar signal generator is found in the frequency domain:

$$S(\omega) = \sum_{n \in Z} 2^{n\gamma} X\left(e^{j2^{n}\omega}\right)\Psi_{0,0}(2^{n}\omega),$$

where X` ($e^{j\omega'}$) denotes the DTFT of the seed signal x(m). Equation and Figure show that the frequency spectrum of x(m) is (periodically) contracted by a factor $2^n$ and directly weights the essential octave band of the corresponding wavelet at scale n, contributing to the frequency spectrum of the self-similar signal for the interval $\left[\frac{\pi}{2^n}, \frac{\pi}{2^{n-T}}\right]$ and its negative frequency mirror.

The type of crossed modulation illustrated emerges in wavelet based fractal modulation: the scaled negative frequency sideband of the seed signal contributes to the spectrum of the self-similar signal in the positive frequencies.

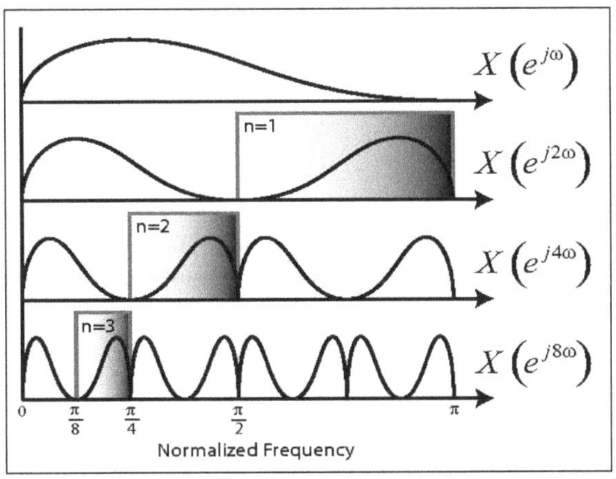


Figure: Frequency spectrum weighting of the wavelet bands (rectangles) by the DTFT of the seed signal.

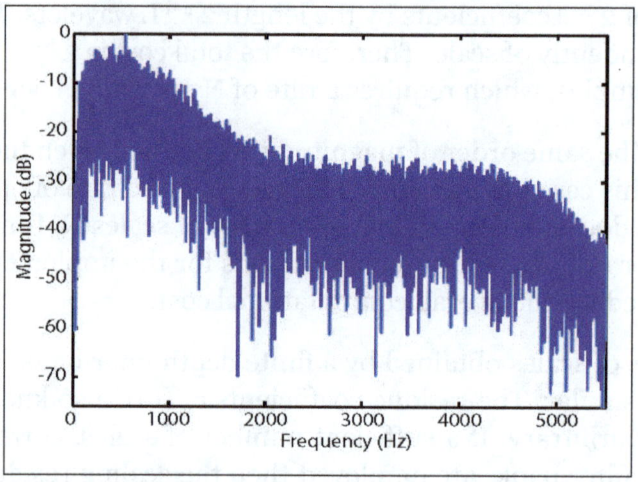


Figure: Frequency spectrum of ocean wave impact sound.

It is shown that the sinc wavelets,

$$\psi_{n,m}(t) = \frac{1}{\sqrt{2^n}} \cos\left(\frac{3\pi}{2}\left(\frac{t}{2^n} - m\right)\right) \operatorname{sinc}\left(\frac{1}{2}\left(\frac{t}{2^n} - m\right)\right),$$

which are based on an ideal rectangular frequency subdivision, form an orthogonal set of wavelets. However, the time localization of the ideal wavelets is very poor and a wide class of regular orthogonal wavelets exist which have a better time-frequency behavior.

## Audio Effects

Interesting and natural sound textures can be produced by means of fractal modulation. The wavelet way to fractal modulation has a low computational cost, which is linear in the number of output samples. The sounds and effects obtained are so different from those produced by conventional, e.g., AM and FM, modulation techniques that an attempt to compare them with fractal modulation is totally meaningless both from a computational and qualitative point of view.

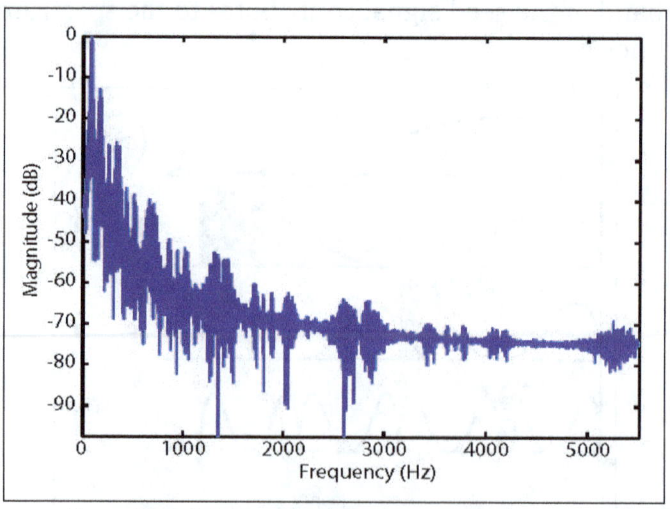


Figure: Frequency spectrum of fractal modulation of the ocean wave impact sound, with $\gamma = 3$ and $N = 7$ scales.

## Dense Textures

In the example of figure we applied the wavelet based fractal modulation scheme to the impact sound of an ocean wave whose frequency spectrum is reported in Figure. For the computation we used N = 7 dyadic scales with Daubechies wavelets of degree 11. The frequency spectrum of the transformed sound shows a highly structured organization, together with an underlying 1/f decay. The modulated sound has a quite apocalyptic flavor, with multivoiced inserts originating from the small scale components and with rhythmic and reverberating patterns originating from the large scale components.

The parameter $\gamma$ controls the overall spectral decay: large values ($\gamma >> 1$) move the spectral centroid to low frequencies, while small values of ($\gamma \approx 1$) tend to produce brighter sounds with flatter spectra, closer to white noise. The parameter $\gamma$ is indeed inversely related to brightness, as it can be seen from equation. For large values of $\gamma$ a much higher weight is associated to the large scale (low frequency) wavelets while for $\gamma < 1$ the small scale (higher frequency) wavelets take the lead.

Sound examples using the ocean wave sound as seed can be found where fractal modulation is applied for various values of the $\gamma$ parameter, showing a corresponding shifting of the spectral center of mass of the sound.

## Sparse Textures

The ocean wave sound has the structure of several fine scale events with an overall swooshing evolution. This type of structure is interesting for the seed to generate complex sounds emulating a large number of concurrent sources. On the other extreme, the application of fractal modulation to sounds with sparser textures, such as the cracking noise of fire, produces an augmented rhythmic texture that lies halfway from the original sound to the sound of dripping water, as it can be appreciated from the corresponding example available.

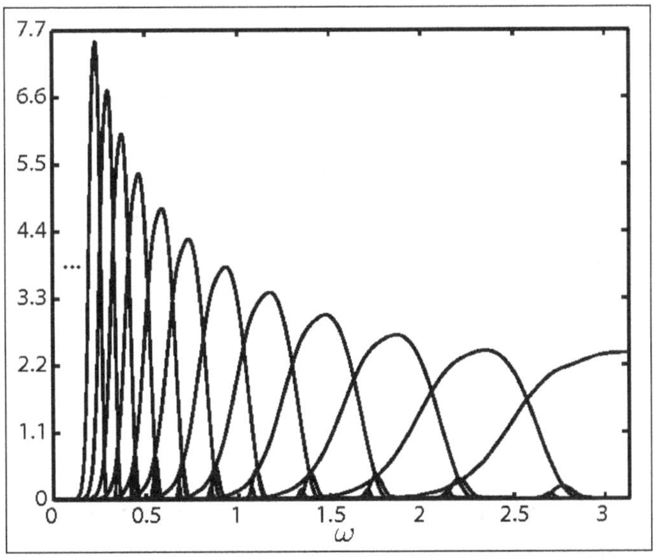


Figure: Magnitude Fourier transform of frequency warped wavelets with 1/3 octave frequency resolution.

In sparse textures it is important or even critical to control the number of scales associated with fractal modulation. Limiting the number of scales achieves sounds where the original rhythmic pattern can still be distinguished, while a multiscale structure is superimposed. At the given URL we show patterns generated using N = 7 and N = 3 scales, which have a different flavor in terms of sharpness of attack and reverberation.

## Extensions

A limitation of the wavelet method for fractal modulation lies in the fact that only dyadic self-similarity with a = 2 can be implemented. As a result, the frequency spectrum is organized by octave bands, which can be too coarse for spectrum modeling. However, orthogonal wavelets having rational scale factor a have been devised based on rational sampling rate filter banks. Alternately, arbitrary scale wavelets, have been defined based on iterated frequency warping schemes, which are suitable for richer fractal modulation.

Frequency warping can also be used in order to modify the perceived pitch, if any, of the sounds generated by fractal modulation. In an example that can be found, frequency warping is employed in order to play a short melody with the fractal modulated ocean wave, which renders the acoustics of a stadium-like crowd singing in an unorganized or spontaneous choir. An efficient approximated frequency warping algorithm can be found in, which allows for real-time computation.

## Dynamic Fractal Modulation

The fractal modulation parameter γ can be dynamically and interactively changed by suitably scaling the seed signal. Since the sampling rate is different for each wavelet component, suitably down sampled amplitude envelopes must be considered. However as the low-frequency components are generated by coefficient sequences at a very low sampling rate, excessive down sampling can result in audible distortion of the time envelopes. A more costly solution for higher fidelity time modulation of the γ parameter can be realized by amplitude scaling the synthesized wavelet components at the same sampling rate as the output signal. In the overlap-add method this can be achieved by the pre-computed shifted wavelets by the proper time-varying factor 2 n γ(k), where γ(k) is the time varying fractal modulation parameter. Interesting brightness transition effects can be realized in his fashion.

## Fractal Modulated Low Frequency Oscillator

Modulated fractal noise has been employed in the Fractal Additive Synthesis technique in order to model the 1/f behavior of the harmonic partials in the sounds of natural instruments. There, the seed signals are given by independent colored noise sources for each wavelet scale. Parallel wavelet synthesis structures are cascaded with a cosine modulated filter bank (MDCT) in order to synthesize the partials, where each wavelet section controls a single sideband of a partial. The same scheme but with deterministic signals replacing the colored noise sources can be employed in order to generate self-similar pseudo-periodic sounds.

Using only the lower frequency fraction of the MDCT, one can generate low-frequency pseudo-sinusoids with fractal fluctuations. These signals with a pseudo-random behavior can prove interesting as control functions of audio effects such as phaser, flanger, chorus, vibrato, e.g., by replacing the traditional LFO with a fractal LFO, whose oscillations are where the main sinusoidal behavior, controlled by the scaling residue sequence $r_N$ (m) of Figure, is not included in the signal in order to better display the fluctuations. The introduction of a time-varying fractal modulation parameter allows us to dynamically control the depth of the effect, ranging from pure sinusoidal to extremely erratic behavior.

## Pulse Modulation Methods

Pulse modulation is a type of modulation in which the signal is transmitted in the form of pulses. It can be used to transmit analogue information. In pulse modulation, continuous signals are sampled at regular intervals.

Pulse modulation can be classified into two major types:

- Analogue: Indication of sample amplitude is infinitely variable.
- Digital: Indicates sample amplitude at the nearest predetermined level.

A block diagram showing the basic classification of modulation techniques is given below.

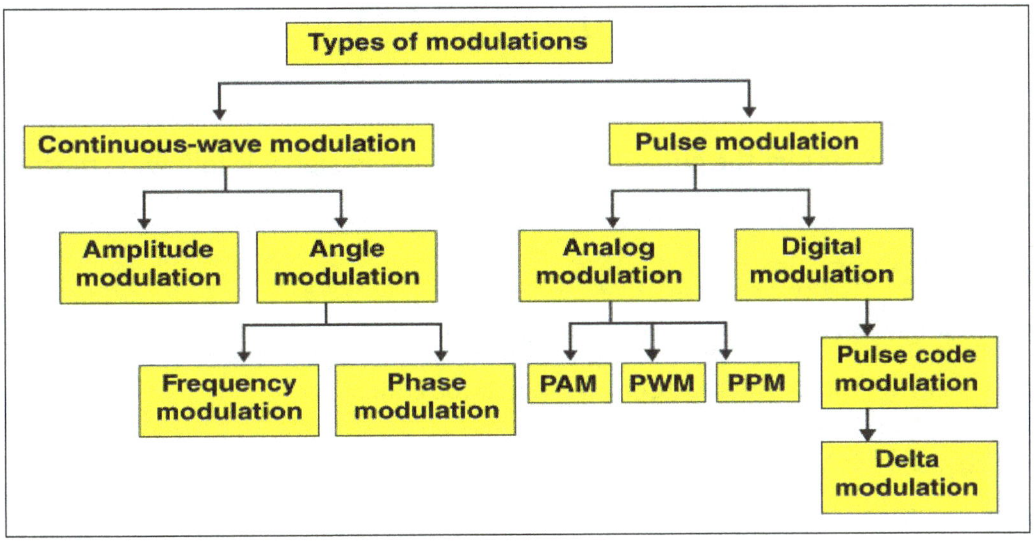


## Types of Pulse Modulation

Let's us look at some of the different types of pulse modulation.

## Pulse Amplitude Modulation (PAM)

It is the simplest form of Pulse Modulation. In this type of modulation, each sample is made proportional to the amplitude of the signal at the instant of sampling. The PAM signal follows the amplitude of the original signal, as the signal traces out the path of the whole wave. Here a signal which is sampled at Nyquist rate can be reconstructed by passing it through an efficient Low Pass Filter (LPF) with exact cutoff frequency. It is very easy to generate and demodulate PAM. This technique transmits the data by encoding in the amplitude of a series of signal pulses.

There are two types of PAM.

- Single Polarity PAM: A fixed DC level is added to the signal so that the signal is always positive.

- Double Polarity PAM: Here the pulses are both positive and negative.

PAM is illustrated in the figure below.

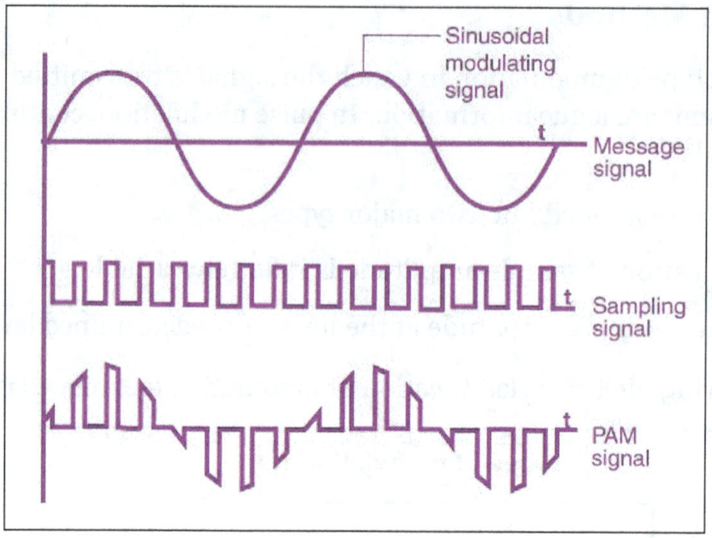


From the figure, it is clear that the pulse amplitude modulated signal is following the amplitude of the message signal.

### Advantages of PAM

- Both Modulation and demodulation are simple.
- Easy construction of transmitter and receiver circuits.

### Disadvantages of PAM

- Large bandwidth is required for transmission.
- More noise.
- Here the amplitude is varying. Therefore, the power required will be more.

### Applications of PAM

- Mainly used in Ethernet communication.
- Many microcontrollers use this technique in order to generate control signals.
- It is used in Photobiology.
- It acts as an electronic driver for LED circuits.

### Pulse Time Modulation (PTM)

Here the pulses will have the same amplitude. However, one of their timing characteristics is made proportional to the amplitude of the sampled signal. This variable characteristic can be either frequency, position or width. This way pulse time modulation can be classified into three types.

## Pulse Width Modulation

Pulse Width Modulation is also known as pulse duration modulation (PDM). Here, as the name suggests, the width of the pulse is varied in proportional to the amplitude of the signal. Since the width is changing, the power loss can be reduced when compared to PAM signals.

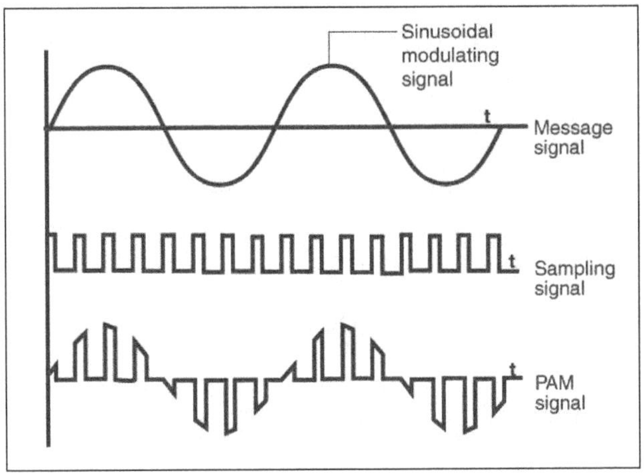


From the figure, it is clear that the amplitude of the signal is constant. Amplitude limiters are used for this. Since clipping of amplitude at desired levels take place, this modulation produces less noise.

## Advantages of PTM

- Low power consumption.
- It has an efficiency of about 90 per cent.
- Noise interference is less.
- High power handling capacity.

## Disadvantages of PTM

- The circuit is more complex.
- Voltage spikes can be seen.
- The system is expensive as it uses semiconductor devices.
- Switching losses will be more due to high PWM frequency.

## Applications of PTM

- Used in encoding purposes in the telecommunication system.
- Used to control brightness in a smart lighting system.
- Helps to prevent overheating in LED's while maintaining it's brightness.
- Used in audio and video amplifiers.

## Pulse Position Modulation

In this type of modulation, both the amplitude and width of the pulse are kept constant. We vary the position of each pulse with reference to a particular pulse. Here a single pulse is transmitted with the required number of phase shifts. So we can say that pulse position modulation is an analogue modulation scheme where the amplitude and width of the pulse are kept constant, while the position of the pulse with respect to the position of a reference pulse is varied according to the instantaneous value of message signal.

PPM can be obtained from PWM. This is done by getting rid of leading-edge and bodies of PWM pulses. The main advantage of pulse position modulation is that it requires constant transmitter power output while the major disadvantage is that it depends upon transmitter-receiver synchronisation. The waveforms of PPM are given below:

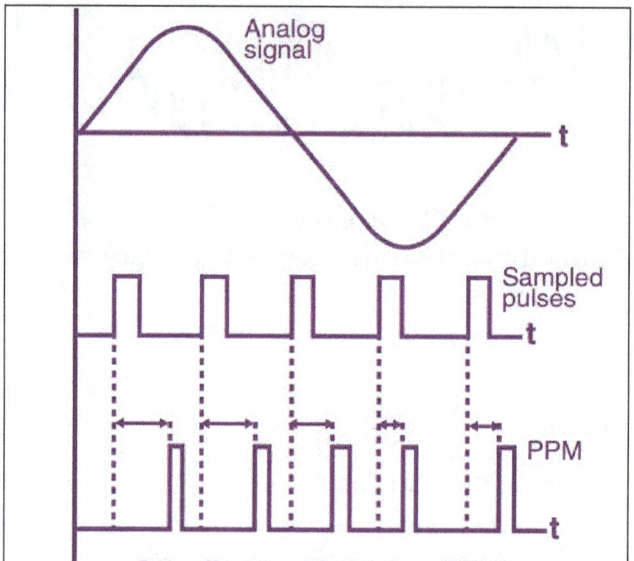


Pulse Position Modulation (PPM)

However, the third type, pulse-frequency modulation is not used.

### Advantages of PPM

- As it has constant amplitude noise interference is less.
- We can easily separate signal from a noisy signal.
- Among all three types, it has the most power efficiency.
- Requires less power when compared to pulse amplitude modulation.

### Disadvantages of PPM

- The system is highly complex.
- The system requires more bandwidth.

## Applications of PPM

- It is used in the air traffic control system and telecommunication systems.
- Remote controlled cars, planes, trains use pulse code modulations.
- It is used to compress data and hence it is used for storage.

## Pulse Code Modulation (PCM)

This type of modulation is different from all modulations learnt so far. It is clear from the block diagram given at the top that it is a type of digital modulation. That is the signals here are sampled and sent in pulse form. A common feature among other techniques is that pulse code modulation also uses sampling technique. In this case, instead of sending a pulse train which is capable of continuously varying parameters, this type of generator produces a series of numbers or digits. Each digit in it represents the appropriate length of the sample at a particular instant. Basic block diagram of its realisation is given below:

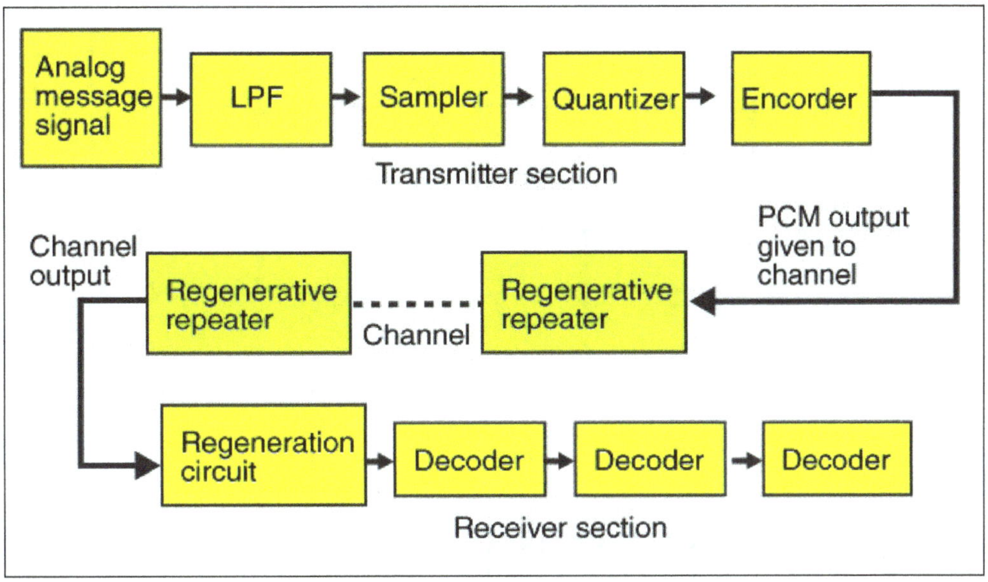


## Advantages of PCM

- It is mainly used in long distant communication.
- Transmitter efficiency is more.
- It has higher noise immunity when compared to other methods.

## Disadvantages of PCM

- More bandwidth is required when compared to analogue systems.
- In this method encoding, decoding and quantisation of the circuit have to be done. This makes it more complex.

### Applications of PCM

- It is used in the satellite transmission system.
- It is also used in space communication.
- Used in Telephony.
- One of the recent applications is the compact disc.

### Delta Modulation

In this type of modulation, the sampling rate is very high. Here, stepsize after quantisation is of the smaller value. In this method, the quantisation design is very simple. Here the bit rate can be designed by the user. Block diagram of a delta modulator is given below:

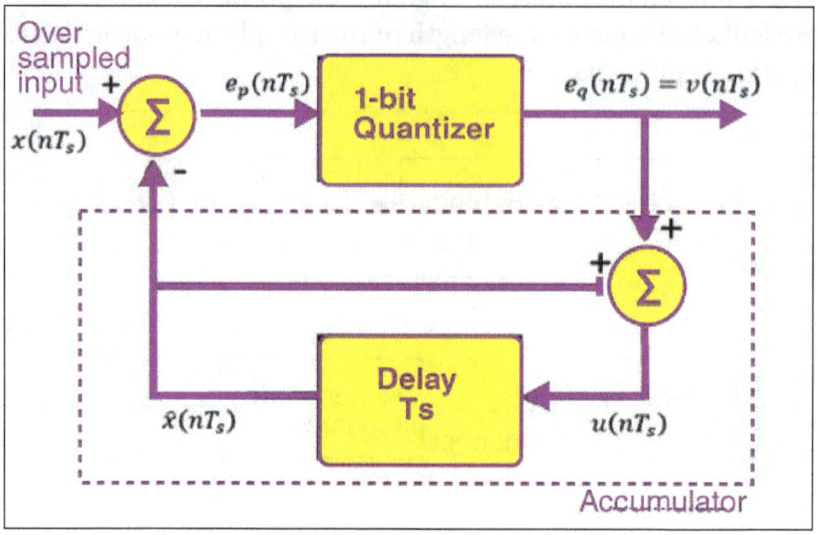


## Antennas and Propagation

An antenna is an electrical conductor or a system of conductors that radiates/collects (transmits or receives) electromagnetic energy into/from space. An idealized isotropic antenna radiates equally in all directions. The directivity of an antenna is captured by its beam width; it is the angle within which power radiated is at least half of that in the most preferred direction (that is, half of the maximum power). Antenna gain is the power output, in a particular direction, compared to that produced in any direction by an idealized omnidirectional antenna. If f is the carrier frequency and $A_e$ is the effective area of the antenna, then the antenna gain is given by,

$$G = \frac{4\pi A_e f^2}{c^2}$$

where c is the speed of light. The effective area of an antenna depends on the size and shape of the antenna. The effective area of an idealized isotropic antenna is 1, by definition.

## Propagation Modes

Wireless transmissions propagate in three modes: ground-wave, sky-wave, and line-of-sight. Ground wave propagation follows the contour of the earth, while sky wave propagation uses reflection by both earth and ionosphere. Finally line of sight propagation requires the transmitting and receiving antennas to be within line of sight of each other. Which of these propagation modes dominates depends on the frequency of the underlying signal.

Examples of ground wave and sky wave communication are AM radio and international broadcasts such as BBC. Above 30 MHz, neither ground wave nor sky wave propagation operates and the communication is through line of sight.

If h is the height of a transmitting (resp., receiving) antenna in meters, then the distance to the receiver (resp., transmitter) for line-of-sight transmission should be at most,

$$d = 3.57\sqrt{h} \text{ kms}$$

This can be proved using elementary geometry. Since microwaves are bent or refracted by the atmosphere, the "effective" line of sight is, in fact, larger than the true line of sight. We introduce an adjustment factor K to capture the refraction effect and obtain,

$$d = 3.57\sqrt{Kh} \text{ kms}$$

If the two antenna have heights h1 and h2 meters, then the distance between them for LOS propagation should be at most,

$$3.57\sqrt{K}\left(\sqrt{h_1} + \sqrt{h_2}\right) kms$$

## Transmission Limitations

There are four classes of limitations that affect electromagnetic wave transmissions.

## Attenuation

The strength of signal falls with distance over transmission medium. The extent of attenuation is a function of distance, transmission medium, as well as the frequency of the underlying transmission. Even in free space, with no other impairment, the transmitted signal attenuates over distance simply because the signal is being spread over a larger and larger area. For an ideal isotropic antenna, the ratio between the transmitted power Pt and the received power Pr for a separation distance of d is given by,

$$\frac{P_r}{P_t} = \frac{(4\pi d)^2}{\lambda^2} = \frac{(4\pi fd)^2}{c^2}$$

where λ, f, and c are the wavelength, frequency, and the speed of the signal.

One can express the above ratio in terms of antenna gains and effective areas of the antennas by applying the appropriate formulae.

## Distortion

Since signals at different frequencies attenuate to different extents, as the above formula indicates, a signal comprising of components over a range of frequencies gets distorted; i.e., the shape of the received signal changes. A standard method of resolving this problem (and recovering the original shape) is to amplify higher frequencies and thus equalize attenuation over a band of frequencies.

## Dispersion

Dispersion is the phenomenon of spreading of a burst of electromagnetic energy during propagation. It is especially prevalent in wireline transmissions such as an optical fiber. Bursts of data sent in rapid succession tend to merge due to dispersion. The longer the length of the wire, the more severe is the effect of dispersion. The effect of dispersion is to limit the product R × L, where R is the data rate and L is distance.

## Noise

The most pervasive form of noise is thermal noise, which is often modeled using an additive Gaussian model. Thermal noise is due to thermal agitation of electrons and is uniformly distributed across the frequency spectrum. The thermal noise to be found in 1 Hz of bandwidth is,

$$No = kT\ W/Hz,$$

where T is the absolute temperature of the medium (conductor) and k is Boltzmann's constant (1.3803×10−23 Joules/Kelvin). The amount of thermal noise in B Hz of bandwidth is simply NoB. Other forms of noise include intermodulation noise (caused by signals produced at frequencies that are sums or differences of carrier frequencies), crosstalk (interference between two signals), and impulse noise (irregular pulses of high energy caused by external electromagnetic disturbances). While an impulse noise may not have a significant impact on analog data, it has a noticeable effect on digital data, causing burst errors.

The effect of noise on the achievable information rate is best captured by Shannon's capacity theorem which we derived last week, which relates the capacity to the signal-to-noise ratio. A more convenient form of this equation is in terms a different ratio, referred to as the Eb/No term. Here Eb refers to the energy per bit of the signal and No is the noise power density introduced above. In terms of the signal power P and the information rate R, we can write Eb as P/R. Noting that N = NoB, we obtain Eb/No = (P/N)(B/R), where B is the bandwidth. Thus, the Eb/No value needed to achieve the capacity of a channel must satisfy the following equation:

$$C = B\log_2\left(1 + \frac{E_b}{N_0}\frac{C}{B}\right) \Rightarrow \frac{E_b}{N+0} = \frac{B}{C}(2^{C/B} - 1)$$

For signal modulation techniques, the bit-error rate is often expressed as a function of the Eb/No ratio.

## Fading

Fading refers to the variation of the signal strength with respect to time/distance and is widely prevalent in wireless transmissions. The most common causes of fading in the wireless environment are multipath propagation and mobility (of objects as well as the communicating devices).

## Multipath Propagation

In wireless media, signals propagate using three principles: reflection, scattering, and diffraction. Reflection occurs when the signal encounters a large solid surface, whose size is much larger than the wavelength of the signal, e.g., a solid wall. Diffraction occurs when the signal encounters an edge or a corner, whose size is larger than the wavelength of the signal, e.g., an edge of a wall. Finally, scattering occurs when the signal encounters small objects of size smaller than the wavelength of the signal. One consequence of multipath propagation is that multiple copies of a signal propagation along multiple different paths, and arrive at any point at different times. So the signal received at a point is not only affected by the inherent noise, distortion, attenuation, and dispersion in the channel but also the interaction of signals propagated along multiple paths.

## Delay Spread

Suppose we transmit a probing pulse from a location and measure the received signal at the recepient location as a function of time. The signal power of the received signal spreads over time due to multipath propagation. The delay spread is determined by the density function of the resulting spread of the delay over time. Average delay spread and root mean square delay spread are two parameters that can be calculated.

## Doppler Spread

This is a measure of spectral broadening caused by the time rate of change of the mobile radio channel. It is caused by either relative motion between the mobile and base station or by movement of objects in the channel. When a pure sinusoid of frequency f is transmitted the received signal spectrum, the Doppler spectrum, is the range [f – fd, f|fd], where fd is referred to as the Doppler spread, given by:

$$f_d = \frac{v}{\lambda} \cos \theta$$

where v is the relative velocity of the mobile and θ is the angle formed by the direction of the motion and that of the signal, and λ is the wavelength of the carrier.

The relative size of the delay spread, as compared to the symbol period, has an impact on whether the effect of fading is uniform over all frequencies or varies across frequencies in the spectrum of the channel. Flat fading occurs when the bandwidth of the signal is less than the "coherence bandwidth" of the channel, or delay spread is less than the symbol period. Frequency selective fading occurs when the bandwidth of the signal is greater than the "coherence bandwidth" of the the channel, or the delay spread is greater than symbol period.

There are several fading models that capture different multipath propagation characteristics. Rayleigh fading models a worst-case scenario in which no path dominates (in fact, a LOS path may not even exist). Rician fading assumes that one of the paths, usually LOS, dominates all the other paths.

When the velocity of the mobile is high, the Doppler spread is high, and the the resulting channel variations are faster than that of the baseband signal; this is referred to as fast fading. When channel variations are slower than the baseband signal variations, then the resulting fading is referred to as slow fading.

## Wireless Transmission

Wireless transmission is a form of unguided media. Wireless communication involves no physical link established between two or more devices, communicating wirelessly. Wireless signals are spread over in the air and are received and interpreted by appropriate antennas.

When an antenna is attached to electrical circuit of a computer or wireless device, it converts the digital data into wireless signals and spread all over within its frequency range. The receptor on the other end receives these signals and converts them back to digital data. A little part of electromagnetic spectrum can be used for wireless transmission.

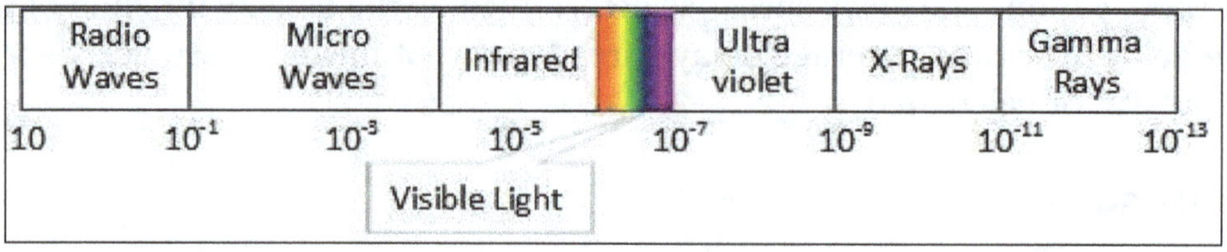


### Radio Transmission

Radio frequency is easier to generate and because of its large wavelength it can penetrate through walls and structures alike.Radio waves can have wavelength from 1 mm – 100,000 km and have frequency ranging from 3 Hz (Extremely Low Frequency) to 300 GHz (Extremely High Frequency). Radio frequencies are sub-divided into six bands.

Radio waves at lower frequencies can travel through walls whereas higher RF can travel in straight line and bounce back.The power of low frequency waves decreases sharply as they cover long distance. High frequency radio waves have more power. Lower frequencies such as VLF, LF, MF bands can travel on the ground up to 1000 kilometers, over the earth's surface.

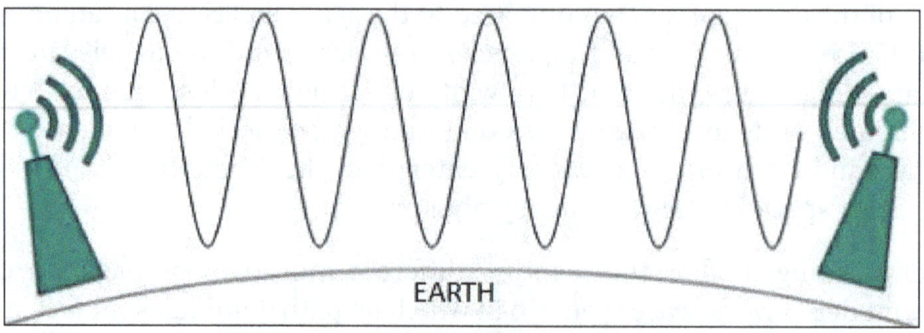

Radio waves of high frequencies are prone to be absorbed by rain and other obstacles. They use Ionosphere of earth atmosphere. High frequency radio waves such as HF and VHF bands are spread upwards. When they reach Ionosphere, they are refracted back to the earth.

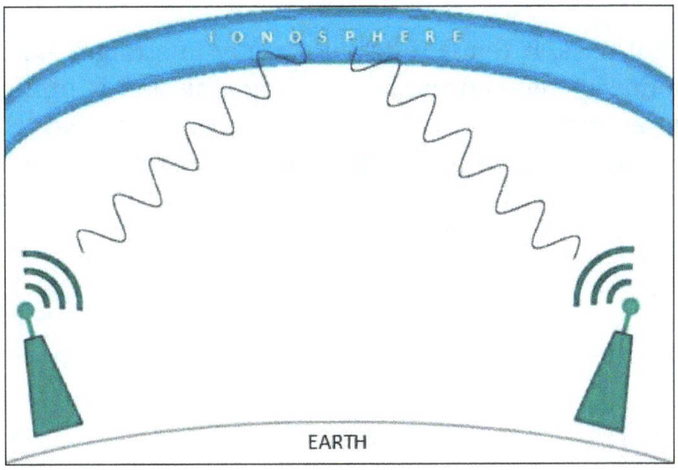


## Microwave Transmission

Electromagnetic waves above 100 MHz tend to travel in a straight line and signals over them can be sent by beaming those waves towards one particular station. Because Microwaves travels in straight lines, both sender and receiver must be aligned to be strictly in line-of-sight.

Microwaves can have wavelength ranging from 1 mm – 1 meter and frequency ranging from 300 MHz to 300 GHz.

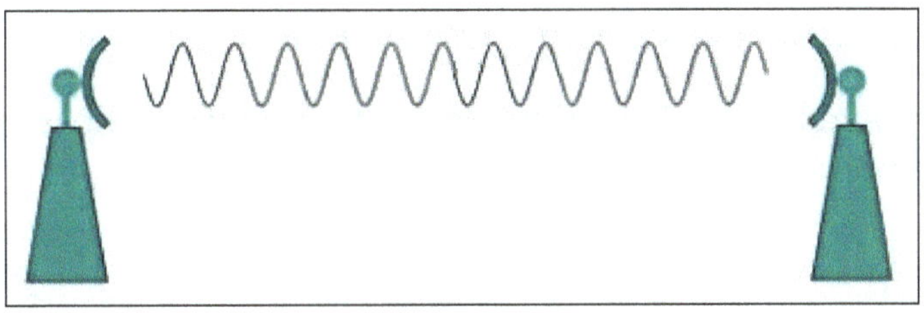

Microwave antennas concentrate the waves making a beam of it. As shown in picture above, multiple antennas can be aligned to reach farther. Microwaves have higher frequencies and do not penetrate wall like obstacles. Microwave transmission depends highly upon the weather conditions and the frequency it is using.

## Infrared Transmission

Infrared wave lies in between visible light spectrum and microwaves. It has wavelength of 700-nm to 1-mm and frequency ranges from 300-GHz to 430-THz.

Infrared wave is used for very short range communication purposes such as television and it's remote. Infrared travels in a straight line hence it is directional by nature. Because of high frequency range, Infrared cannot cross wall-like obstacles.

## Light Transmission

Highest most electromagnetic spectrum which can be used for data transmission is light or optical signaling. This is achieved by means of LASER.

Because of frequency light uses, it tends to travel strictly in straight line.Hence the sender and receiver must be in the line-of-sight. Because laser transmission is unidirectional, at both ends of communication the laser and the photo-detector needs to be installed. Laser beam is generally 1mm wide hence it is a work of precision to align two far receptors each pointing to lasers source.

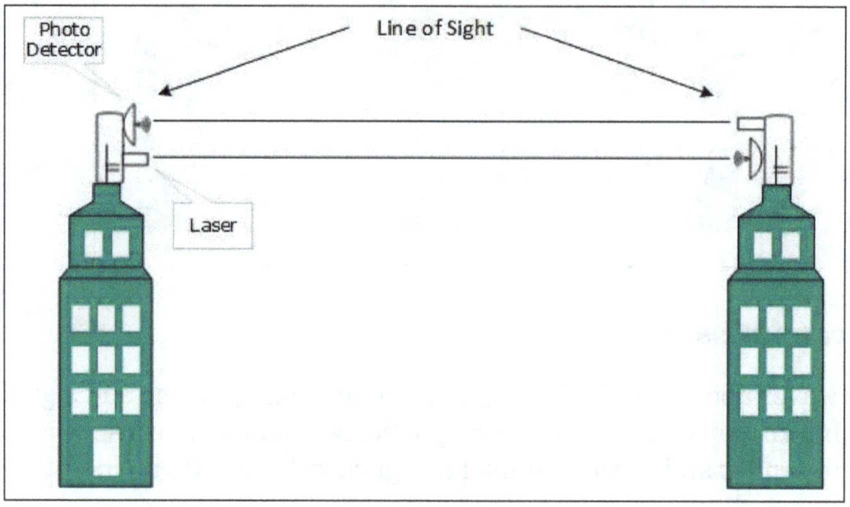


Laser works as Tx (transmitter) and photo-detectors works as Rx (receiver). Lasers cannot penetrate obstacles such as walls, rain, and thick fog. Additionally, laser beam is distorted by wind, atmosphere temperature, or variation in temperature in the path.

Laser is safe for data transmission as it is very difficult to tap 1mm wide laser without interrupting the communication channel.

## References


- Mobile-communication-introduction: javatpoint.com, Retrieved 12, March 2020
- Mobile-communication-protocols, communication-technologies: tutorialspoint.com, Retrieved 25, February 2020
- Principles-of-communication-modulation, principles-of-communication: tutorialspoint.com, Retrieved 18, June 2020
- Analog-modulation, modulation-techniques, PulseBlasterDDS-300: Retrieved 10, April 2020
- Digital-modulation-different-types-and-their-differences: elprocus.com, Retrieved 05, January 2020
- Phase-modulation-shift-keying-what-is-psk, radio, modulation: electronics-notes.com, Retrieved 11, July 2020
- Frequency-shift-keying-fsk-working-applications: elprocus.com, Retrieved 03, May 2020
- Wireless-transmission, data-communication-computer-network: electronics-notes.com, Retrieved 19, August 2020

3

# Medium Access Control

A medium access control is a network data transfer policy responsible for the flow control and data transmission through a wired, optical or wireless transmission medium. It ensures non-collision and eases transfer of data packets between two terminals. FDMA, TMA, CDMA and SDMA are some of the medium access schemes used in medium access control. This chapter touches on the subject of medium access control and medium access schemes in detail.

The medium access control (MAC) sublayer is closely associated with the physical layer and defines the means by which the physical channel (medium) may be accessed. It coordinates the attempts to seize a shared channel by multiple MAC entities to avoid or reduce the collisions in it. The MAC sublayer commonly provides a limited form of error control, especially for any header information that defines the MAC level destination and higher layer access mechanism.

Ethernet (IEEE 802.3) is a prime example of a shared medium with a defined MAC sublayer functionality. The shared medium in Ethernet has traditionally consisted of a coaxial cable into which multiple entities were "tapped." Although this topology still applies conceptually, a hub and spoke medium is now typically used, in which the earlier coaxial cable has been physically collapsed into a hub device.

As a contention medium, Ethernet defines how devices sense a channel for its availability, wait when it is busy, seize the channel when it becomes available, and backs-off for a random length of time following a collision with another simultaneously transmitting device. On a shared channel, such as Ethernet, only a single entity can transmit at a time or frames will be garbled.

Not all shared channels involve contention. A prime example of a shared medium without contention is token ring (IEEE 802.5), in which control of the channel is rotated between the devices sharing the channel in a deterministic round-robin manner. Conceptually, control of the channel is given to the entity currently possessing a "token." If the device has nothing to transmit, it passes the token to the next device attached to the topological "ring."

IEEE-defined MAC sublayer addresses are six bytes long and permanently assigned to each device, typically called a network interface card. The IEEE administers the assignment of these addresses in blocks to manufacturers to ensure the global uniqueness that the MAC sublayer protocols rely on for "plug on play" network setup. Each manufacturer must ensure individual device identifier uniqueness within their assigned block.

A MAC PDU consists of a MAC header, zero or more MAC SDUs, zero or more MAC control elements, and optional padding. The MAC header and MAC SDUs may have variable sizes. As shown in Figure, a MAC PDU header consists of one or more MAC PDU subheaders where each subheader corresponds to a MAC SDU, a MAC control element, or padding. A MAC PDU subheader consists of the six header fields R/R/E/LCID/F/Length except for the last subheader in the MAC PDU and for fixed-size MAC control elements. The last subheader in the MAC PDU and subheaders for fixed-size MAC control elements as well as the subheader for padding consist of the four header fields R/R/E/LCID. Note that a MAC PDU is a bit string that is byte-aligned.

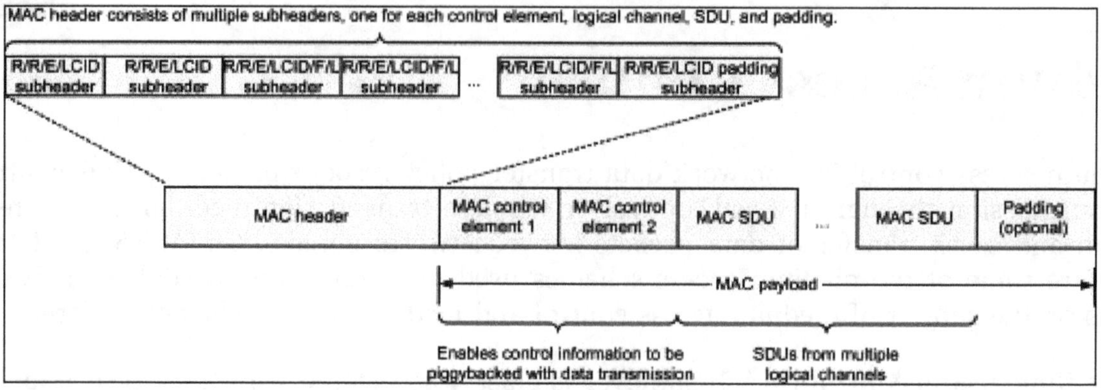

Figure: MAC PDU formation.

The structure of the MAC PDU used on UL-SCH and DL-SCH is shown in Figure. Each MAC PDU corresponds to one transport block and consists of a header, control elements, logical channel SDUs (corresponding to RLC PDUs), and optional padding. MAC control elements are used to piggyback control information, such as buffer status or power headroom reports. The MAC header consists of several subheaders. There is one subheader for each MAC control element and each MAC SDU in the PDU. In addition, there can be a subheader to specify padding. For the control element, the subheader identifies the specific control element being sent. For the MAC SDUs, the logical channel number and the size of the SDU are specified. All subheaders are grouped at the start of the MAC PDU in the MAC header. These are followed by the control elements, which are then followed by the MAC SDUs. Any padding occurs at the end of the PDU unless only one or two bytes of padding are required. Then one or two bytes of fixed-size padding subheaders can be included before the subheader corresponding to the first MAC SDU.

The subheader formats for MAC PDUs are shown in Figure. There are two formats: [R/R/E/LCID] which is used for fixed-length MAC SDUs and MAC control elements; and [R/R/E/LCID/F/Length] that is used for variable length MAC SDUs. The size of the first format is one byte. The size of the second format is either two or three bytes. In each case, the two R bits are reserved bits and are set to "0." The E bit is used to indicate if there are more subheader fields in the MAC header. A value of "1" denotes that another subheader would follow. A value of "0" indicates that a control element or MAC SDU starts at the next byte. The LCID field indicates either the logical channel number associated with the MAC SDU or, in the case of a control element, the specific type of MAC control element. The length field can be either 7 or 15 bits long and specifies the length of the MAC SDU in bytes. The size of the Length field is determined by the F field, which is "0" for 7 bits (128 bytes or less), or "1" for 15 bits.

Medium access control or multiple access control (MAC) has been one of the main topics under discussion in relation to cellular networks in which a base station (BS) communicates to multiple users in downlink and multiple users communicate to a BS in uplink in each cell. In classical circuit-switched networks with voice communication as their main application, such as 1G and 2G cellular networks, the channel resource between the BS and multiple users is divided into a fixed number of orthogonal channels, and these orthogonal channels are assigned to users who request voice communication. When all channels are occupied by users there are no more channels left for additional users and additional users requiring communication service experience communication resource shortage called blocking. In such schemes, the number of orthogonal channels is determined to maintain a certain probability of blocking with consideration of the average number of users in the cell and their activity factor. Examples of such schemes include frequency-division multiple access (FDMA) in 1G and time-division multiple access (TDMA) and code-division multiple access (CDMA) in 2G.

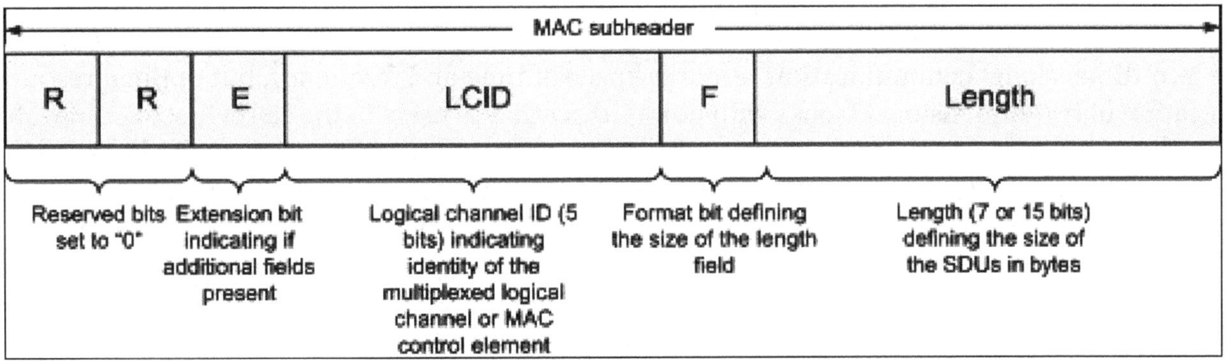


Figure: Structure of a MAC header.

With the pervasive use of the Internet in late 1990s, efficient wireless packet data communication was incorporated into cellular networks. Since then, wireless packet data communication has been crucial in cellular networks. The design of wireless packet data cellular networks is inherently different from that of circuit-switched cellular networks. First, the arrival of wireless packets, e.g., emails and image files, for each user is bursty. Second, most applications for packet communication require much less stringent delay constraints. These two aspects of wireless packet data combined with the properties of wireless channels enabled a new paradigm for MAC design in cellular networks. Traditionally, wireless channel fading or fluctuation due to mobility was considered to be a detrimental effect that should be compensated for by power control to provide a constant quality communication link for each voice user. In wireless packet data communication, however, channel fading is exploited rather than compensated for to yield good overall system performance. Suppose that the communication resource such as time is divided into blocks in time domain and each resource block is assigned to one of many packet data users, as in early single user (SU) selection scheduling methods, and that each user experiences independent channel fading over time blocks. Then, there is a high probability that the wireless channel of one of the users is in good condition when there are many users in the cell. By assigning each communication resource block to the user with the best channel condition at that time block, it is possible to achieve good overall system performance. In this case, the channel fading is not compensated for but exploited for efficient system performance. The diversity gain provided by the selection of a user or users with a good channel condition among the many users with independent channel fading is called multiuser (MU) diversity and the process of dynamically selecting some users for each communication resource block among the many active users in the cell based on their channel conditions is called user scheduling. MU diversity and user scheduling are the key aspects of MAC design for cellular packet networks.

MU diversity gain varies with channel environments and scheduling methods, and thus various scheduling algorithms have been proposed under different channel and system environments over the last two decades. In the early packet data cellular networks, user scheduling was performed in one dimension, for example, the TDMA-based downlink of the IS-856 cellular standard. In this one-dimensional case, the communication resource such as time is divided into one-dimensional blocks and a single user is scheduled to each resource block. Thus, the design of a scheduler is a rather simple problem. For example, the proportional fair (PF) scheduling algorithm applied to the one-dimensional case optimizes the system data rate and the fairness among the users, and it was adopted in IS-856. However, as orthogonal frequency-division multiplexing/multiple access (OFDM/OFDMA) was adopted as the physical-layer transmission technology in IEEE 802.16 (WiMAX), IEEE 802.20 (MBWA), and 3GPP LTE, more flexible resource allocation was possible over the two-dimensional communication resource space of time and frequency, but optimal resource allocation of two-dimensional blocks with power allocation to users in the cell is more complicated than the one-dimensional case. Various scheduling and resource allocation methods have been proposed for OFDM/OFDMA downlink and uplink networks under various criteria. Furthermore, the MIMO technology was incorporated to OFDM/OFDMA for 4G cellular networks. With the MIMO technology, multiple users can simultaneously be supported at the same time-frequency resource block based on spatial-division multiple access (SDMA). Thus, in 4G cellular networks adopting MIMO-OFDM user scheduling can be performed in the three-dimensional resource space of time, frequency, and space, and optimal user scheduling in the three-dimensional resource allocation is a far more difficult problem than that of the one- or two-dimensional systems. Thus, many practical scheduling algorithms simplify the problem by using the capability of multiple data stream transmission provided by MIMO to transmit multiple data streams to one user. In this case, for each resource block SU MIMO is implemented between the BS and the selected user for the resource block and scheduling in MIMO-OFDM networks reduces to the previous two-dimensional problem. However, it was shown that the sum rate of SU-MIMO-based MIMO downlink scheduling achieves only a small fraction of the sum rate capacity of a MIMO broadcast channel (BC) when the number of transmit antennas at the BS is larger than that of the receive antennas at each user. This is mainly because the extra number of transmit antennas at the BS is used to increase the power gain rather than the multiplexing gain associated with multiple antennas. On the contrary, scheduling based on MU-MIMO, where multiple data streams provided by MIMO are assigned to multiple users at the same time-frequency resource block based on beamforming, can achieve a large fraction of the sum rate capacity of a MIMO BC by attaining the full multiplexing gain. It has been shown that when the number of active users in the cell grows without bound, the optimal sum rate capacity of a MIMO BC obtained by dirty paper coding (DPC) can be achieved asymptotically by some smart MU scheduling. Specifically, both the sum rate capacity of a MIMO BC and the sum rates of smart MU-MIMO scheduling algorithms scale as for fixed M under independent Rayleigh fading assumption, where M and K are the numbers of transmit antennas and active users in the cell, respectively.

Recently, the MIMO technology has evolved to massive MIMO, which is considered to be one of the key technologies for 5G. In massive MIMO, large-scale antenna arrays are used at BSs to harness the gain of MIMO beyond the small-scale MIMO of four or eight antennas at the BS used in 4G. With massive MIMO it is possible to attain significant capacity increases and energy efficiency. In rich scattering environments, such as in sub-6 GHz radio bands, the use of a large number of antennas results in asymptotic orthogonality among users' channel vectors and thus simple linear processing, such as maximum ratio transmission (MRT) in downlink and maximum ratio combining (MRC) in uplink, can be used to remove inter-user interference. Furthermore, the use of a large number of antennas also results in channel hardening, which means that small-scale channel fading vanishes due to the law of large numbers. This channel hardening effect simplifies resource allocation and power control. Specifically, frequency-domain user scheduling may not be necessary because all subcarriers of OFDM/OFDMA have almost the same channel gain due to the channel hardening. Thus, the whole bandwidth can be assigned to each scheduled user separated by MIMO in the spatial domain. In this case, the user scheduling problem focuses on how to select multiple users simultaneously supported in the spatial domain at each scheduling time interval. Massive MIMO generally requires a large amount of feedback from each user due to the large dimension of channel vectors and thus it is important to devise scheduling algorithms having reasonable performance with a relatively small amount of feedback based on partial channel state information (CSI) feedback from each user. Several scheduling algorithms have been proposed and analyzed for massive MIMO in this respect.

Another key technology for 5G is the use of mmWave bands for commercial cellular networks. The use of mmWave bands provides wide bandwidth for ultra-high data rates. However, the radio propagation characteristics in mmWave bands are quite different from those in sub-6 GHz bands mostly characterized as rich scattering with the popular Rayleigh or Rician fading channel model. In mmWave bands, the radio propagation experiences severe path loss and there exist very few multiple paths. Such large path losses can be compensated for by very sharp beamforming based on large antenna arrays at the BS and user terminals. Consequently, it is natural to incorporate massive MIMO into wireless communication in mmWave bands. From the perspective of user scheduling, it is important to analyze the performance of conventional MU-MIMO scheduling algorithms in the new mmWave channel environment and devise MU-MIMO scheduling algorithms suitable for the mmWave channel environment.

## Medium Access Scheme

### FDMA

Frequency-division multiple access is the oldest of all multiple access schemes, and was used in the first demand assignment system for satellites (i.e., single channel per carrier (SCPC)) and the first-generation of cellular mobile systems (i.e., the advanced mobile phone system (AMPS)). In frequency-division multiple access (FDMA), the available channel bandwidth is divided into many nonoverlapping frequency bands, where each band is dynamically assigned to a specific user to transmit data. In an FDMA system, signals, while occupying their assigned frequency bands, can be transmitted simultaneously and continuously without interfering with each other. In FDMA, there is a central controller that allocates the frequency band to users, solely based on their needs. This is usually done during the call set up. Once a band is allocated to a user, it then belongs to the user exclusively for the continuous flow of information during the call. To prevent interference, the allocated bands are separated from one another by small guard bands. In other words, FDMA allows the users to transmit simultaneously, but over disjoint frequency bands, a user exploits a fixed portion of the band all the time, as shown in Figure. FDMA is best suited for connection-oriented applications; it is, however, inefficient in terms of utilization of power and bandwidth. If an FDMA channel is not in use by the user, then it sits idle and cannot be used by other users. It has poor spectral efficiency, since guard bands must be employed to avoid overlapping of the adjacent channels, and this in turn reduces the channel capacity.

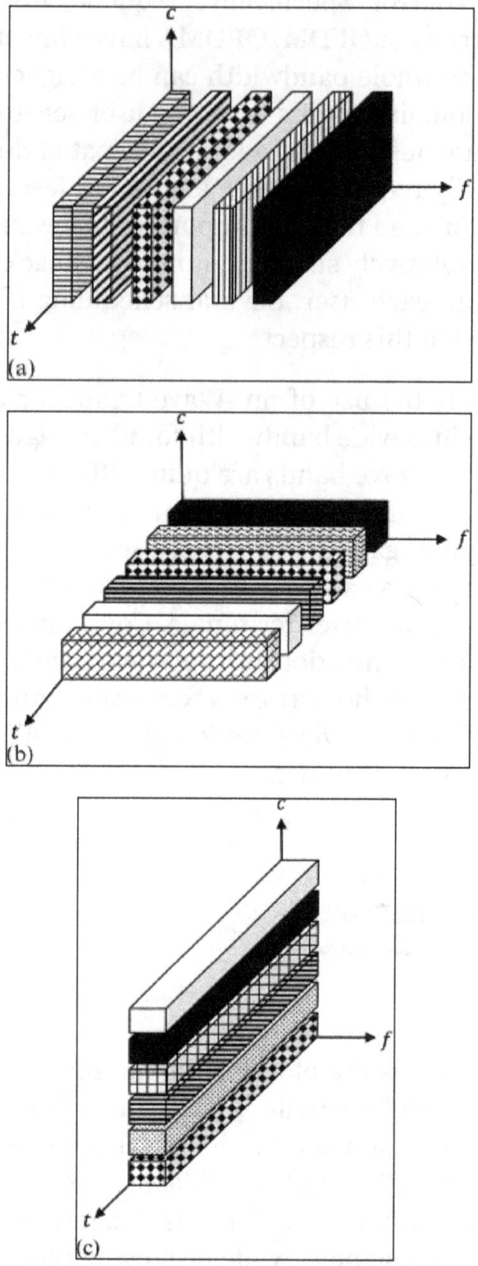


Figure: Multiple access: (a) frequency-division multiple access, (b) time-division multiple access, and (c) code-division multiple access.

FDMA assigns individual channels to individual users. It can be seen from Figure that each user is allocated a unique frequency channel. These channels are assigned on demand to subscribers who request service. Guard bands are maintained between adjacent signal spectra to minimize cross talk between channels. During the period of the call, no other user can share the same frequency band. In frequency division duplex (FDD) systems, the users are assigned a channel as a pair of frequencies; one frequency is used for the upward channel, while the other frequency is used for the downward channel.

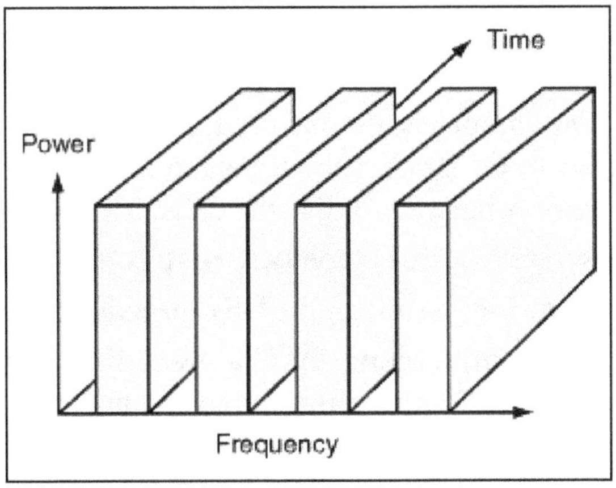


FDMA is used by analog systems, such as AMPS, NMT, or Radiocom 2000. The advantages of FDMA are the following:

- The complexity of FDMA systems is lower when compared to TDMA and CDMA systems, though this is changing as digital signal processing methods improve for TDMA and CDMA.
- FDMA it is technically simple to implement.
- A capacity increase can be obtained by reducing the information bit rate and using efficient digital codes.
- Since FDMA is a continuous transmission scheme, fewer bits are needed for overhead purposes as compared to TDMA.

FDMA uses frequency allocation among a set of earth stations to share a series of frequencies. The transponder bandwidth is divided into a series of nonoverlapping frequency slots, each of which constitutes an access channel. Because the nonlinearity of the satellite transponder causes intermodulation products, nonlinear signal power transfer, and intelligible crosstalk, the TWTA cannot be operated to full capacity. Therefore, the power efficiency of an FDMA system is decreased because of the need to decrease the power of the TWTA. Other potential problems in FDMA are as follows:

- During uplink, RF power coordination is needed to prevent the suppression of weaker signal by stronger signals.
- Low flexibility; need to slowly change preassigned traffic patterns.
- Carrying capacity of traffic is relatively low.

## Single-Carrier Frequency Division Multiple Access

SC-FDMA has been used as the uplink multiple-access scheme in LTE systems. The use of SC-FDMA was motivated by the fact that a single-carrier system with an OFDMA-type multiple-access would combine the advantages of the two techniques, that is, low PAPR and large coverage. The first SC-FDMA concept was interleaved FDMA (IFDMA), which was based on compression and block repetition of the modulated signal in the time domain. It can be theoretically shown that the spectrum of the compressed and K times repeated signal has the same shape as the original signal, with the difference that it presents exactly K-1 zeros between two data subcarriers. This feature enables us to easily interleave different users in the frequency domain by applying to each user a specific frequency shift, or equivalently, by multiplying the time-domain sequence by a specific phase ramp. In addition, similar to OFDMA, robustness to inter-cell interference can be achieved by coordinating resource allocation between adjacent cells. The same waveform can be obtained in the frequency domain, if discrete Fourier transform matrix $P = \left[p_{k,n}\right]; p_{k,n} = \exp(j2\pi kn / M)$ is used as the precoding matrix in Fig, resulting in DFT-precoded OFDMA, which is mathematically identical to IFDMA in a distributed scenario. The precoding operation Ps is equivalent to an point DFT operation. With a subcarrier mapping matrix Ω as given in Section, the spectrum of the distributed DFT-precoded OFDMA signal is identical to the IFDMA signal spectrum, thus it corresponds to the same waveform. This is also called DFT-spread OFDM. The two techniques are different implementations of SC-FDMA. The advantage of DFT-precoded OFDMA is in its more flexible structure. While IFDMA imposes a distributed signal structure, DFT-precoded OFDMA allows the use of an appropriate subcarrier mapping matrix $\Omega$. Localized variants of implementation or channel-dependent mappings are also possible. A pulse-shaping filter can be further applied in the frequency domain, with a lower complexity than the time-domain filtering. Note that in frequency-selective channel scenarios, interference may occur among the M elements of each data block. This degradation, which is more important in a distributed subcarrier mapping, also impacts Walsh–Hadamard precoded OFDMA.

## Disadvantages

- The system architecture based on FDMA was implemented in first-generation analog systems such as advanced mobile phone system (AMPS) or total access communication system (TACS). The improvement in capacity depends on operation at a reduced signal-to-interference (S/I) ratio. But the narrowband digital approach gives only limited advantages in this regard so that modest capacity improvements could be expected from the allocated spectrum.
- The maximum bit-rate per channel is fixed and small, inhibiting the flexibility in bit-rate capability that may be a requirement for computer file transfer in some applications in the future.
- Inefficient use of spectrum, in FDMA if a channel is not in use, it remains idle and cannot be used to enhance the system capacity.
- Crosstalk arising from adjacent channel interference is produced by nonlinear effects.

### TDMA

Time Division Multiple Access (TDMA) is a digital cellular telephone communication technology. It facilitates many users to share the same frequency without interference. Its technology divides a signal into different timeslots, and increases the data carrying capacity.

Time Division Multiple Access (TDMA) is a complex technology, because it requires an accurate synchronization between the transmitter and the receiver. TDMA is used in digital mobile radio systems. The individual mobile stations cyclically assign a frequency for the exclusive use of a time interval.

In most of the cases, the entire system bandwidth for an interval of time is not assigned to a station. However, the frequency of the system is divided into sub-bands, and TDMA is used for the multiple access in each sub-band. Sub-bands are known as carrier frequencies. The mobile system that uses this technique is referred as the multi-carrier systems.

In the following example, the frequency band has been shared by three users. Each user is assigned definite timeslots to send and receive data. In this example, user 'B' sends after user 'A,' and user 'C' sends thereafter. In this way, the peak power becomes a problem and larger by the burst communication.

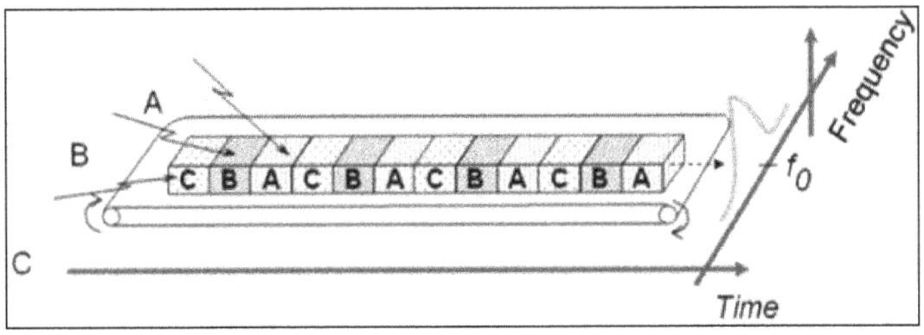


## FDMA and TDMA

This is a multi-carrier TDMA system. A 25 MHz frequency range holds 124 single chains (carrier frequencies 200) bandwidth of each kHz; each of these frequency channels contains 8 TDMA conversation channels. Thus, the sequence of timeslots and frequencies assigned to a mobile station is the physical channels of a TDMA system. In each timeslot, the mobile station transmits a data packet.

The period of time assigned to a timeslot for a mobile station also determines the number of TDMA channels on a carrier frequency. The period of timeslots are combined in a so-called TDMA frame. TDMA signal transmitted on a carrier frequency usually requires more bandwidth than FDMA signal. Due to the use of multiple times, the gross data rate should be even higher.

## Advantages of TDMA

Here is a list of few notable advantages of TDMA:

- Permits flexible rates (i.e. several slots can be assigned to a user, for example, each time interval translates 32Kbps, a user is assigned two 64 Kbps slots per frame).
- Can withstand gusty or variable bit rate traffic. Number of slots allocated to a user can be changed frame by frame (for example, two slots in the frame 1, three slots in the frame 2, one slot in the frame 3, frame 0 of the notches 4, etc.).
- No guard band required for the wideband system.
- No narrowband filter required for the wideband system.

## Disadvantages of TDMA

The disadvantages of TDMA are as follow:

- High data rates of broadband systems require complex equalization.

- Due to the burst mode, a large number of additional bits are required for synchronization and supervision.
- Call time is needed in each slot to accommodate time to inaccuracies (due to clock instability).
- Electronics operating at high bit rates increase energy consumption.
- Complex signal processing is required to synchronize within short slots.

## CDMA

Code Division Multiple Access, CDMA was an innovative use of direct sequence spread spectrum technology used to provide a multiple access scheme for mobile telecommunications and other wireless systems.

CDMA used the property of DSSS that unless the transmitter and receiver used the same spreading code for both ends of the process, the signal could not be decoded and in this way it was able to provide a means of enabling a variety of different users to use the same channel to access a base station without mutual interference. In this way, using CDMA different users were allocated different codes rather than different slots, channels, etc.

Figure: Typical 3G mobile phone that uses CDMA.

CDMA is based around a form of transmission known as Direct Sequence Spread Spectrum. The CDMA history can be directly linked back to the 1940s when this form of transmission was first envisaged. As electronics technology improved, it started to be used for covert military transmissions in view of the facts that the transmissions look like noise, it is difficult to decipher without the knowledge of the right codes, and furthermore it is difficult to jam.

With the revolution in cellular telecommunications that occurred in the 1980s a then little know company named Qualcomm working on DSSS transmissions started to look at this as the basis for a cellular telecommunications multiple access scheme - CDMA - code division multiple access.

The concept of CDMA had to proved in the field and accordingly Qualcomm was joined by US network operators Nynex and Ameritech to develop the first experimental CDMA system. Later the team was expanded as Motorola and AT&T (now Lucent) joined to bring their resources to speed development.

As a result this it was possible to start writing a specification for CDMA in 1990. With the support of the Cellular Telecommunications Industry Association (CTIA) and the Telecommunications Industry Association (TIA) a standards group was set up. This group then published the standard for the first CDMA system in the form of IS-95, resulting in the formal publication of IS-95-A in 1995.

The first CDMA system was launched in September 1995 by Hutchison Telephone Co. Ltd. in Hong Kong and SK Telecom in Korea soon followed along with networks in the USA.

This was only one cellular telecommunications system, although it was the first. Its development lead on to the CDMA2000 series of standards.

The use of CDMA did not stop with CDMA2000 as it became necessary to evolve the GSM standard so that it could carry data and provide significant improvements in terms of spectrum use efficiency. Accordingly CDMA, in the form of Wideband CDMA (WCDMA) was adopted for this standard.

## Key Elements of CDMA

CDMA is a form of spread spectrum transmission technology based around a scheme called direct sequence spread spectrum.

By proving different users with different spreading codes, the different users are able to utilise the same frequency channel, whilst still being able to individually communicate with the base station.

The use of CDMA has been likened to being in a room where there are many people speaking different languages, and being able to understand someone speaking in your language despite the high level of noise. A CDMA signal will be able to be decoded when the receiver uses the same code as used for the transmission despite the presence of other signals with different codes being used on the same channel. CDMA has a number of distinguishing features that are key to spread spectrum transmission technologies:

- Use of wide bandwidth: CDMA, like other spread spectrum technologies uses a wider bandwidth than would otherwise be needed for the transmission of the data. This results in a number of advantages including an increased immunity to interference or jamming, and multiple user access.
- Spreading codes used: In order to achieve the increased bandwidth, the data is spread by use of a code which is independent of the data.
- Level of security: In order to receive the data, the receiver must have a knowledge of the spreading code, without this it is not possible to decipher the transmitted data, and this gives a measure of security.
- Multiple access: The use of the spreading codes which are independent for each user along with synchronous reception allow multiple users to access the same channel simultaneously.

## CDMA Technology Advantages

The use of CDMA offers several advantages and it is for this reason that CDMA technology has been adopted for many 3G cellular telecommunications systems.

Improvement in handover / handoff: Using CDMA it is possible for a terminal to communicate with two base stations at once. As a result, the old link only needs to be broken when the new one is firmly established. This provides significant improvements in terms of the reliability of handover / handoff from one base station to another.

Improvement in capacity: One of the chief claims for CDMA is that it gave significant improvements in network capacity. Original expectations for some of the proponents of CDMA technology were for some very significant improvements, although in reality these were somewhat exaggerated over what real world experience found:

- 18 fold increase in capacity when compared to AMPS (1G technology used in USA).
- 6 fold increase in capacity when compared to US TDMA (2G technology used in USA) - similar increases were also claimed over GSM.
- In reality the original expectations were not fulfilled although increases of a factor of about two were seen when compared to US TDMA and GSM. This in itself was a significant improvement.
- CDMA< code division multiple access provided the multiple access technology for all the major 3G cellular telecommunications systems. As such it gave a significant step forwards over the TDMA / FDMA schemes used for 2G, enabling data transmission in a way that had not been previously possible.

## SDMA

Spatial division multiple access (SDMA) is a channel access method used in mobile communication systems which reuses the same set of cell phone frequencies in a given service area.

PACKET-SWITCHED networks are seen as the generic network architecture for future wireless systems that must support mixed traffic in an efficient manner. In the presence of multimedia traffic and environmental changes, it is required that a medium access control (MAC) can effectively adapt to changing channel conditions and traffic requirements. Many innovative MAC protocols have been proposed in the past decade. Most schemes, however, inherit some architectural features of wire-line systems and treat "layers" as uncoupled entities. For example, data link/media-access protocols have been frequently designed and analyzed without sufficient regard to design details at the physical layer. One of the key resources in wireless communication is the spatial diversity provided by antenna arrays. While inherently rich, the spatial resource is highly irregular and, thus, is difficult to deal with at the MAC level. For this reason exploitation of the spatial diversity has been limited to the physical layer using techniques such as spatial beamforming. However, if the MAC protocol is designed without adaptability to the channel conditions, it has to be designed for the worst-case scenario and as a result, there will be too much signal level margin and a significant waste of network resources.

The need for coordinated MAC and physical layers in a system architecture is pronounced in antenna array networks. Only limited results are available to date, one of which is the so-called space-division multiple-access (SDMA) scheme that attempts to multiply the throughput of a wireless local area network (WLAN) by incorporating spatial channels. The SDMA scheme is particularly suitable for fixed wireless networks where the spatial characteristics are relatively stable. For a system with M antenna elements, the underlying idea is to divide each time-division multiple-access (TDMA) slot into M space slots so that the total number of noninterfering slots is increased by M fold.

Though intuitively promising, an obvious flaw of this scheme is that spatial channels are rarely orthogonal in practice. If multiple terminals are assigned to one time slot without considering their spatial characteristics, the one with an unfavorable spatial configuration will experience significant throughput disadvantages. Since the effectiveness of spatial separation depends on the basestation array responses (often referred to as the spatial signatures) of all co-slot terminals, the instantaneous signal-to-noise ratio (SNR) of beamforming outputs can vary dramatically. This problem is investigated by Ward and Compton where collision due to unresolvable packets (when the arrival angles of multiple packets are within a threshold $\Theta_r$ ) is accounted for. Although the analysis is somewhat simplified, since in the presence of multipath the capability of reception algorithms may not be a sole function of the arrival angles, the study reveals the limitations of the basic SDMA scheme.

A fundamental solution to the above problem is a "channelaware" MAC that controls the traffic based on the spatial characteristics of the terminals. In SDMA in particular, the performance of the system can be enhanced with spatial signature based scheduling (e.g., assigning the "most orthogonal" terminals to the same time slot to increase the traffic throughput). Such a MAC treatment allows a system to exploit the spatial diversity in an efficient manner with fixed-complexity physical layer processing. Several dynamic scheduling algorithms are proposed and studied in. All of them assume a static homogeneous traffic with terminals transmitting packets at all times. It is shown that under steady traffic, the scheduling algorithm provides a minimum performance gap between the average and the worst case throughput among the terminals, indicating a maximum capacity improvement in worst case limited applications. In most real scenarios, however, the network traffic is bursty and heterogeneous due to mixture of voice, video and data, each with different quality-of-service (QoS) parameters. In this case, the MAC protocol has to take the traffic dynamics and QoS requirements into account. Based on the existing scheduling algorithm, we propose an enhanced scheme for multimedia packet networks. In this protocol, not only the terminals' spatial signatures, but also their QoS parameters are considered in order to optimized the system performance. Simulation shows that a substantial improvement can be obtained than the best effort scheme in which all traffics are treated homogeneously.

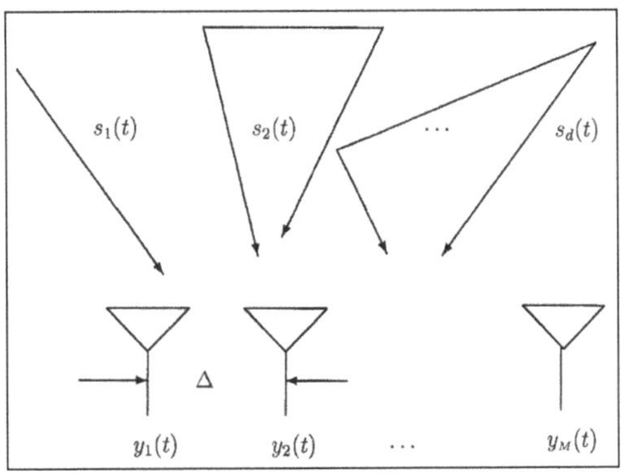


Figure: A typical array system.

## Antenna Array Packet Networks

### Spatial Beamforming

Antenna array processing has been an active topic in wireless communications for more than a decade. Through controlling its beam pattern, an antenna array can improve some key operation parameters such as the signal-to-interference-plus-noise-ratio (SINR) over single antenna wireless system. Fig. depicts a typical antenna array system where the spatial diversity is exploited for multiple access communications. In the absence of noise, the response of an M-element antenna array to a narrowband source s(t) can be written as:

$$y(t) \overset{def}{=} [y_1(t)y_2(t)....yM(t)]^T = as(t)$$

where $(.)^T$ denotes transposition, and $a=[a(1)a(2)...a(M)]^T$ is the array response vector that captures the spatial characteristics (direction-of-arrivals, number of multipath reflections, and attenuation) associated with the terminal. In some literature the array response vector is referred to as the spatial signature. When d terminals communicate simultaneously with the basestation, the total output of the antenna array is given by:

$$y(t) = \sum_{k=1}^{d} a_k s_k(t) + n(t) = As(t) + n(t)$$

$$A = [a_1,....,a_d], = s(t)[s_1(t),...,Sd(t)]^T$$

$$n(t) = [n_1(t), n_2(t),...,n_M(T)]^T$$

Where n(t) is an additive noise vector and A is defined as the array manifold whose columns are the spatial signatures, we assume 1 all signals have unit power and $\| a_k \| = 1$ (i.e., perfect power control) and 2) the noise is i.i.d. with strength $\sigma \frac{2}{n}$. Under these assumptions the covariance matrix of the array output has the form of,

$$R_{yy} = E\{y(t)y^H(t)\} = AA^H + \sigma \frac{2}{n} L$$

In most practical situations, spatial signatures of different terminals are different, allowing the basestation to decouple superimposed signals through spatial separation. To retrieve individual signals, e.g., from the antenna outputs, outputs of the antenna array are weighted and summed with a set of weight coefficients,

$$\hat{S}_i(t) = \sum_{m=1}^{M} w^*_i(m) ym(t)$$

with $\{w_i(m)\}$ so designed to constructively combine the signal of interest and destructively combine the interference-plus-noise. This processing is referred to as spatial beamforming. The optimum weight vector $w_k = [w_k(1)...w_k(M)]$ that minimizes the mean squared error (MSE) of the signal estimate is given by,

$$w_k = R_{yy}^{-1} a_k$$

in which case the output MSE and SINR, respectively, of the kth signal estimate are,

$$MSE_k = 1 - a_I^H R_{yy}^{-1} a_i;$$

It can be shown that through spatial beamforming, up to M co-channel terminals can be perfectly separated in the absence of noise. The capability of simultaneous communications with multiple terminals using an antenna array makes SMDA very attractive to resource critical wireless applications.

## SDMA/TDMA Packet Networks

Before we review the concept of SDMA/TDMA, let us describe the packet radio network under consideration. The slotted packet network consists of K radio terminals actively communicating with a basestation. Information is transmitted in blocks of symbols called packets. The time axis is divided into time frames of fixed length. Each frame is evenly divided into L time slots through which multiple packets can be transmitted. Each terminal transmits packets in one out of the L slots at a fixed data rate.

Assume that the basestation has total control of the network traffic. When a terminal has packets to transmit, it places an admission-request packet through a reserved request slot, from which the basestation obtains the spatial signature and traffic information of the terminal. The basestation then assigns certain time slots to the terminal depending on the request and the traffic situation. The focus is on how to schedule packet transmissions after the terminals are admitted. We do not emphasize access initiation in this work.

In most practical situations. SDMA/TDMA accommodates these terminals in both the space and the time domain. Fig. illustrates how this objective is achieved in a 4-slot, 10-terminal SDMA/TDMA network. Built upon the slotted TDMA network, the SDMA expands the capacity by allowing multiple (up to M) terminals in each slot, and within each slot spatial beamforming is performed to acquire packets from multiple terminals. In essence, SDMA adds another dimension to the spectrum resource by expanding each slot into M space slots (vertical axis), translating into an M-fold increase of system throughput in an ideal scenario.

In reality however, the ability of capturing multiple packets depends critically on the spatial configuration of the co-slot terminals. Intuitively, by assigning "most-orthogonal" terminals to the same time slot, the worst case and average performance of the system can be improved. This motivates us to design efficient scheduling algorithms that separates "aligned" terminals so that the spatial resource is efficiently exploited. Note that the scheduling criteria depends heavily on the system configuration. Thus there is no single "optimal" solution for all the systems. In the following, we focus on a TDMA-based single-cell system with perfect power control. We try to develop scheduling algorithms to accommodate packets with fixed size into the time slots to improve the system performance.

### Scheduling Algorithms

Assume that the transmission data rate of each terminal is fixed. Then the terminal has to operate above certain SINR threshold $SINR_{min}$ to avoid excessive packet errors. The threshold depends on the coding and modulation scheme and the packet PER requirement. In the scheduling process, a terminal is allocated to a time slot as long as the resulting SINR after beamforming is above the threshold $SINR_{min}$.

$$SINR \geq SINR_{min}$$

Shad et al. studied several scheduling algorithms to improve the frame capacity. Among all algorithms studied, the Best Fit algorithm has the best performance.

The best-fit scheduling algorithm first pick a terminal randomly from the terminal pool (to ensure the fairness among terminals). The algorithm tests the current unallocated terminal by calculating the SINRs for all co-slot terminals in the current time slot as if the candidate terminal were to be added. The SINR margins of co-slot terminals are calculated by subtracting $SINR_{min}$ from the SINR they experienced. Assume there are d terminals in current slot, The SINR margin of the th terminal in this slot is: $SINR_{margin}(i)=SINR(i)-SINR_{min}$. The the SINR margin of this slot is. $SINR_{margin}=\min\{SINR_{margin}(i)\}$The algorithm would test all the time slots and insert the terminal into the time slot with the largest SINR margin provide that the margin is greater than zero. Otherwise, the terminal is ignored. The algorithm then randomly picks the next terminal. The process continues until all the terminals are tested.

The best-fit scheme implicitly assumes that the traffic is steady, i.e., all terminals have packets to transmit at all times. In real applications the number of terminals changes every time a terminal becomes active/inactive. Under these scenarios it may not be reasonable to perform a full-scale scheduling on all terminals every time when there is a change in network traffic. While optimization for variable traffic patterns is an important topic that requires further investigation, one straightforward strategy is partial best-fit scheduling, i.e., allocating only the newly activated terminals based on their spatial signatures. We shall refer to this scheme as "partial-fit" SDMA. Let $\{m_k\} \underset{k=1}{K}$ be the K be the mobile terminals in the system, and $M_l^{f-1} \equiv \{m_{l1} \ldots m_{ld_l}\}$ be the terminal set allocated to the l[th] time slot in the f-1[th] frame. Then, in the f[th] frame, the "partial-fit" algorithm updates the terminal allocation as follows:

- Let $M_l^f = M_l^{f-1}$ for$l$=1,...,L.
- Remove all the expired terminals. If terminal $m_{lk} \in m_l^{f-1}$ finishes it's transmission, let $m_l^f = m_1^f - m_{lk}$, where subtraction removes the terminal from the set.
- Add new terminals. If $m_{f1}, \ldots, m_{fn}$ are newly arrived terminals, then they are added to the existing terminal profile $m_l^f$ ,$l$=1,...,L according to the "best-fit" algorithm described above.

## Performance of Scheduling SDMA

Let $\{a_k\} \overset{K}{k} = 1$ be the spatial signatures of the terminals in the system,

$$A_l=[a_{l1} \ldots a_{ldl}]$$

denote the array manifold of the co-slot terminals in the lth slot. The basestation performs spatial beamforming on the received co-slot signals. According to Poor and Verdu, the residual interference and noise of the MMSE beamforming output has a Gaussian-like characteristic, allowing us to calculate the bit-error rate (BER) based on the output SINR in,

$$P_E = Q(SINR).$$

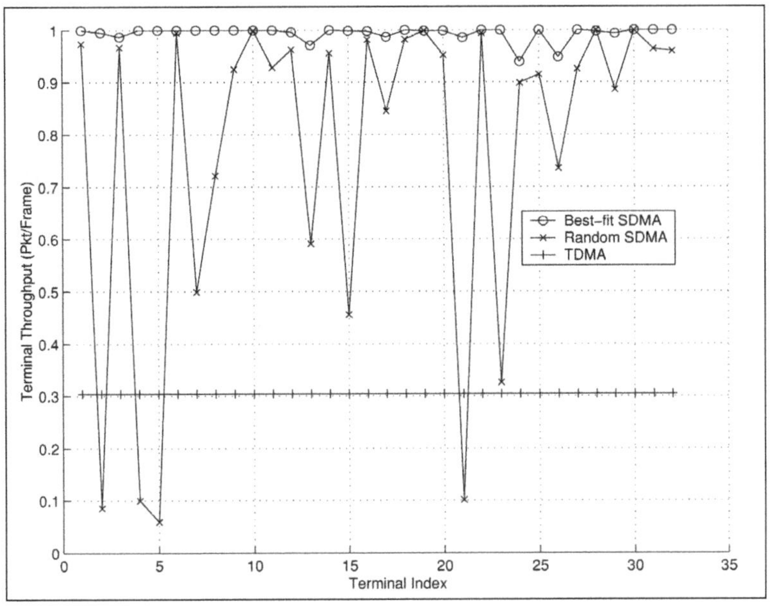


Figure: Throughput improvement of SDMA due to SS-based allocation.

Assuming the error correction capability of a coded packet of N bits is t, i.e., a maximum number of t bits in error is correctable, the packet success probability is,

$$P_k = P(SINR_k)$$
$$= \sum_{n=o}^{t} \binom{N}{n} Q(SINR_k)^n (1 - Q(SINR_k))^{N-n}$$

Where is $SINR_k$ the beamforming output SINR of the kth terminal. The throughput of each terminal in SDMA can be defined accordingly.

Definition 1: The terminal frame throughput $S_k$ is defined as the packet throughput of the kth terminal in a frame. Let be the total number of time slots in a frame. Note that in a given time slot, the kth terminal can transmit at most one packet. Denote $P_{kl}$ the packet success probability of the kth terminal in the lth slot. If the terminal dose not transmit in this slot,. Then, $S_k$ is the summation of all $P_{kl}$,

$$S_k = \sum_{l=1}^{L} Pkl.$$

In cases where the frame length is variable, the average number of packets can be successfully transmitted in a time slot is a better performance measure. The average slot throughput is defined as,

$$T = \frac{1}{L} \sum_{k=1}^{K} S_k.$$

Figure illustrates the performance gain of SDMA by comparing the terminal frame throughput of pure TDMA (single antenna), random SDMA (terminals allocated to slots without regarding to their spatial signatures), and best-fit SDMA/TDMA. We consider a TDMA-based single-cell system, where the basestation is equipped with four antennas. Each TDMA frame has ten time slots, and each time slot accommodates packets with fixed length of 256 b. There are 32 terminals in the system. All terminals are under perfect power control and the spatial signatures are generated as independent Gaussian vectors (normalized to unit power). The SNR is set at 20 dB and SINR threshold is 10 dB. Clearly, SDMA/TDMA is fundamentally better than pure TDMA in that each terminal has four times the chance to transmit successfully. However, the worst case performance of random SDMA could be even worse than pure TDMA due to high possibility of collisions of co-channel terminals. With simple MAC layer planning, the system throughput of best-fit SDMA/TDMA increases substantially across the board. More importantly, the variance of the terminals' throughput becomes very small, one can practically use the average throughput as the design parameter for the system capacity. This is particularly important to wireless applications whose capacity is usually worst case limited.

## Scheduling Protocol for Multimedia Traffic

In the presence of multimedia traffic, different traffic class may have different QoS requirements. For example, the voice traffic is sensitive to delay, but can tolerate certain degree of packet loss. However, the data traffic must not have any missing packet, but can endure longer delay. Therefore, it is beneficial for the MAC protocol to take the specific QoS parameters of each traffic class into account. In the following, we describe a scheduling protocol that significantly increases the performance of a multimedia SDMA network by incorporating the following QoS parameters:

- SINR.
- Packet timeout.
- Packet loss rate.

We consider a single-cell network with one base station and N mobile terminals. Fixed-length packets arrive at the mobiles according to bursty random processes. The packets are buffered at the mobile terminals until they are transmitted to the base station.

When a mobile terminal has packets to transmit, it sends an access packet in the access slot using ALOHA random access. After the basestation successfully receives an access packet from a mobile terminal, it obtains information from the mobile terminal which includes the QoS class of the packets and the spatial signature of the terminal. Based on received information, the basestation then grants certain time slots to the terminal according to a desired packet transmission policy. The transmit permission is broadcasted in the following downlink frame.

The mobile terminal listens to the downlink frame, when it hears the permission, it begins to transmit in the next uplink frame. Each time a mobile terminal transmits a packet, it also includes a Piggybacking Request to indicate whether it has more packets backlogged. The basestation checks the Piggybacking Request, and updates the mobile terminal information accordingly.

Now our focus is on how to design a packet scheduling policy so that the mobile terminals share the limited communications bandwidth in an efficient manner: maximizing throughput and minimizing packet loss. To accommodate the different QoS requirement, a packet scheduler is added to the protocol to schedule packet transmissions in such a way that the spatial resource can be used efficiently. The ensuing sections describe how the scheduler prioritizes and organizes the transmission of packets by taking these criteria into consideration.

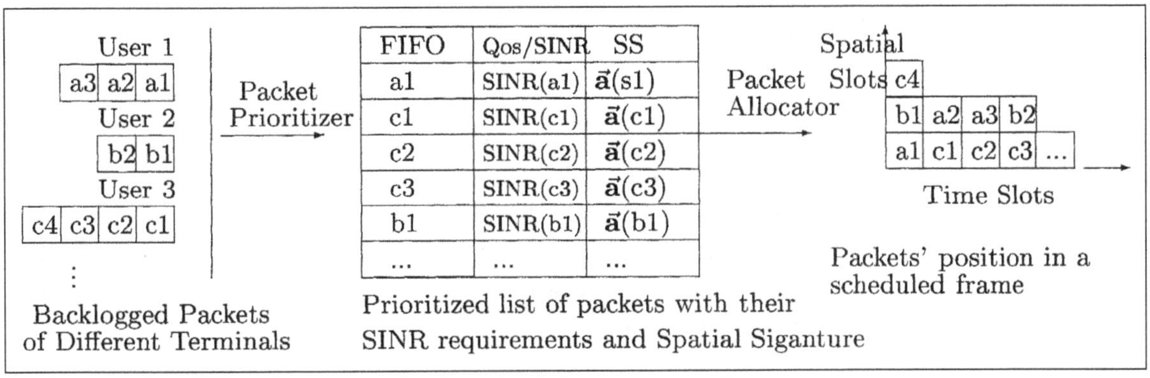


Figure: Packets assigning procedure.

The main objective of the scheduler is to maximize throughput and minimize packet loss. This is achieved in two steps:

- Determine packet priorities.
- Allocate prioritized packets into SDMA/TDMA slots. These tasks are performed by a packet prioritizer and a packet allocator, respectively. In particular, the packet prioritizer minimizes the packet loss, while the packet allocator maximizes the throughput. The basic protocol is illustrated in figure.

## Packet Prioritizer

There are many ways to define packet priorities. Here, we consider the case where each terminal only supports one type of traffic (voice, data, or video). However, terminals can be easily configured to support heterogeneous traffics by maintaining multiple first-in–first-out (FIFO) queues for different traffics. Since the packets in each queue are of the same type, the priority values can be determined based on their time out requirements.

Assume that the kth terminal has $B_K$ backlogged packets in its buffer, with timeout values[t1,t2,...,$t_{BK}$. Ideally, if every packet in the queue is allowed to transmit at a minimal constant rate so that it can finish transmission before timeout, the rate of the ith packet is $r_k(i) = 1/t_i$. The aggregate transmission rate at current frame is thus $r_k = \sum_{i=1}^{Bk} r_k(i)$. It indicates how many slots are required by this terminal at current frame. We can use the aggregate transmission rate as the measurement of priority.

Definition 2: The priority of a packet is defined as the aggregate transmission rate of this packet and the remaining packets backlogged after it. So the priority of the th packet in th terminal queue can be calculated as,

$$P_k(i) = \sum_{j=i}^{Bk} \frac{1}{t_j}.$$

If a packet cannot be transmitted before its time out, it is simply discarded. $P_k(i)$ Is a monotonic decreasing function of. Its value reflects how urgent one packet must be transmitted, because a higher value indicates a higher probability of packet loss. The value also indicates the average transmission rate this terminal should have. If the value is greater than one, it means more than one slot should be assigned to this terminal to avoid possible packet loss. So the priority definition adopted here has the following properties: 1) the priorities of each queues are calculated independently such that it has much lower complexity; 2) the priority based on timeout value gives advantage to the more time sensitive packets (voice, video); 3) the priority values indicate how many slots should be allocated to a terminal in the current frame; and 4) the priority values are calculated based on all buffered packets so that the longer queues which subject to higher packet loss rate can have more advantage. The priority defined here is not optimal because the optimal solution would require to consider all the existing queues jointly. However, given the desirable properties described above, it works well as we will show in the simulations.

After prioritization, the packets are put into a FIFO (an ordered list $L_P$ according to their priority values). Notice that the frame length is $L$. So one terminal can be assigned up to $L$ time slots. So one terminal can have at most $L$ priority values, and it can appear in the list for up to $L$ times. As will be explained next, the list $L_P$ is used by the packet allocator to determine if a terminal should be granted transmission privileges for the next frame, and if so, the number of packets that will be allowed to send. After each frame, the packet prioritizer updates the priority list for the next frame.

## Multimedia Best-Fit Packet Allocator

After prioritization, each ordered packet enters the buffer with an associated SINR/data rate requirement and a spatial signature. The objective of the packet allocator is to arrange packets that have the highest priority into SDMA/TDMA frames so that the throughput in the next frame is maximized. Toward this end, we modify the best-fit algorithm to accommodate the heterogeneous traffic. First of all, the order of t the backlogged packets are determined by the prioritizer. In addition, we allow one terminal to have more than one slot to transmit packets within each frame. The list $L_P$ implicitly contains information of the number of packets that the packet allocator should allow a terminal to send. Furthermore, packets belong to different traffic classes may be assigned to one time slot, as long as their own SINR requirements are all satisfied. Taking these factors into consideration, a QoS-aware SDMA scheduling algorithm is summarized as follows.

Multimedia Best-Fit SDMA/TDMA:

- The algorithm picks packets in the order given by list $L_P$. Each packet is associated with its SINR requirement determined by its QoS class.
- The scheduling starts with the first packet in $L_P$. The algorithm tries to assign the current packet into one of the time slots. The algorithm tests the current unallocated packet by calculating the SINR for all co-slot packets in current time slot as if the candidate packet were to be added. The SINR margins of co-slot packets are calculated by subtracting their SINR requirements from the SINR they experienced. Assume there are d terminals in current slot, The SINR margin of the ith terminal in this slot is: $SINR_{margin}(i) = SINR(i) - SIN_{req}(i)$. The SINR margin of this slot is $SINR_{margin}(i) = \min\{SINR(i)_{margin}(i)\}$. The algorithm would

test all the time slots and insert the current packet into the time slot with largest SINR margin provide that the margin is greater than zero. Otherwise, the packet is ignored.

- The algorithm moves to next packet and continues to search for appropriate slot to insert. The process continues until all packets are tested.

The packet prioritizer orders the packets according to their priority. The packet allocator then searches and allocates packets according to their priority order. The throughput is maximized for the given packet priority order.

## Performance Evaluation

Our analysis so far assumes that all terminals have packets to transmit at all times, which is not always true in practical wireless networks. In this section, we present some performance results for a network with more realistic traffic models. The SDMA/TDMA schemes (random SDMA, best-fit SDMA, partial-fit SDMA and multimedia best-fit SDMA) are studied under homogeneous bursty traffic and mixed voice and data traffic.

## Homogeneous Bursty Traffic

For a network with homogeneous bursty traffic, we assume:

- There are a fixed number of terminals in the system each with its own spatial signature.
- The packets arrive at each terminal according to a bursty traffic model. In this model, the terminal has two alternating states: ON and OFF. During the ON state, packets arrive at the terminal with Poisson distribution. No packet arrives in the OFF period. The length of the ON and OFF periods obey certain distributions.
- The spatial signature of each terminal varies slowly during the ON period, so that all the packets arrived during one period are transmitted with the same spatial signature.

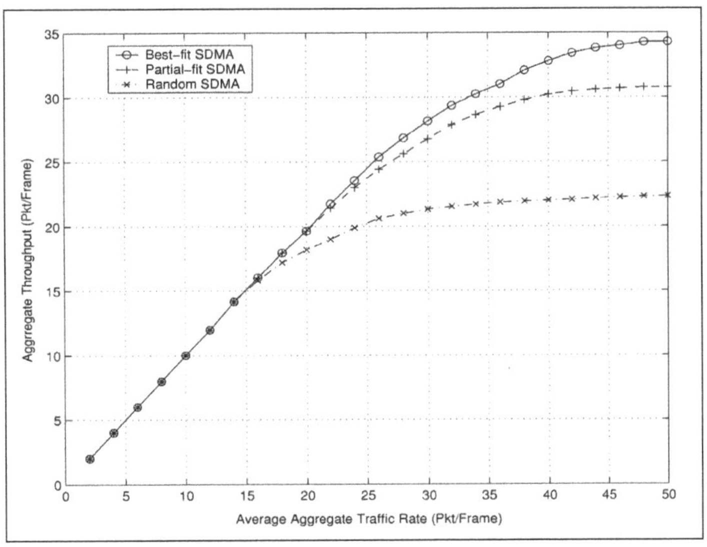


Figure: Throughput with bursty traffic.

The above case presents a realistic WLAN type of scenario most suitable for SDMA. In practice, once a terminal becomes ON, it submits a request to the scheduler and the scheduler marks this terminal as active and estimates its spatial signature. It either randomly assigns one slot to this terminal (random SDMA) or allocates a slot based on its spatial signature (best-fit SDMA).

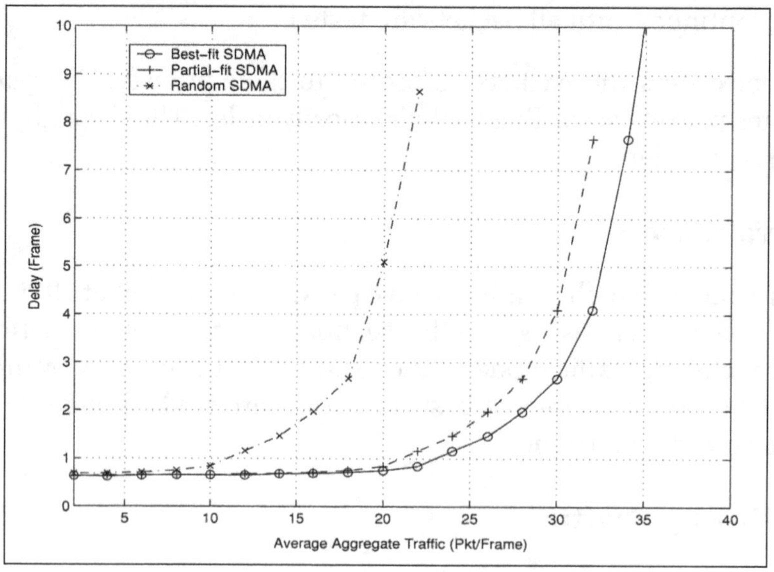


Figure: Delay with bursty traffic.

The terminal, after receiving the scheduling information, begins to transmit the backlogged packets at the given slot. We simulate a basestation with four antennas serving 160 terminals. There are ten time slots per frame. In each slot, the packet length is 256 b. The SNR is set at 20 dB, and the scheduling SINR threshold is 10 dB. The length of the OFF period is exponentially distributed with mean time 100 frames. The ON period has fixed length 1/2 frame. The average aggregate traffic rate $\lambda$ is given by the equation at the bottom of the page.

Figures shown the throughput and delay of the three SDMA schemes as functions of the aggregate traffic rate $\lambda$. For low traffic, the rise in throughput is approximately linear with respect to the traffic rate for all schemes. In this region, there are essentially no collisions due to failure in beamforming acquisition. As the traffic increases, the effect of scheduling becomes more and more evident. The throughput of best-fit SDMA saturates at around 35 packet/frame when the traffic rate is greater than 50 packet/frame, showing a throughput gain of over 50% than that of random SDMA (without scheduling). The improvement in delay characteristics due to scheduling is as significant as that in throughput. As evident from the figure, there is only a small performance gap (about 10%) between the best-fit SDMA (full scheduling) scheme and the partial-fit SDMA (partial scheduling) scheme for the bursty traffic. This is particularly encouraging for practical bursty data applications because of its reduced system overhead:

$$\lambda = \frac{\textit{Traffic Rate during ON period} \times \textit{Number of Users} \times \textit{ON Period}}{\textit{ON Period} + \textit{OFF Period}}$$

Table: Voice traffic parameters.

| Patterns | Mean Value |
|---|---|
| Main Talkspurt | 1.000 |

| Main Gap | 1.350 |
|---|---|
| Minispurt | 0.275 |
| Minigap | 0.050 |

Table: Simulation parameters.

| Parameter | Value |
|---|---|
| Frame Size | 16 msec |
| Slots per Frame $N_p$ | 10 |
| Packet Length | 256 |
| Number of Antennas | 4 |
| SNR | 20db |
| Voice Traffic | 16 Kbps |
| | 2 frames |
| | 9db |
| | 1 percent |
| Data Traffic | 15 Kbps |
| | 200 frames |
| | 12 db |
| | $10^{-6}$ |

## Multimedia Traffic

We now evaluate the performance of the proposed protocol by considering an SDMA system with two types of traffics: voice and data. Each terminal only generates one type of traffic, and there are the same number of voice terminals and data terminals in the system. The voice traffic model is based on the three-state Markov model in. In the model, it is assumed that a speech source generates patterns of talkspurts and gaps as the result of the talking, pausing, and listening behaviors in a conversation. All spurts and gaps have independent exponentially distributed durations. The values for the speech activity model is listed in Table. The data packet length and packet arrival interval are modeled as exponentially distributed. The mean data unit size is 30 kB. The simulation parameters are listed in table.

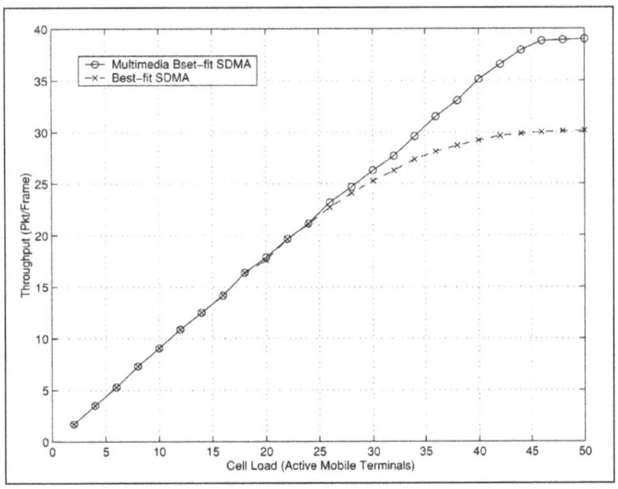


Figure: Throughput per frame vs. cell loads.

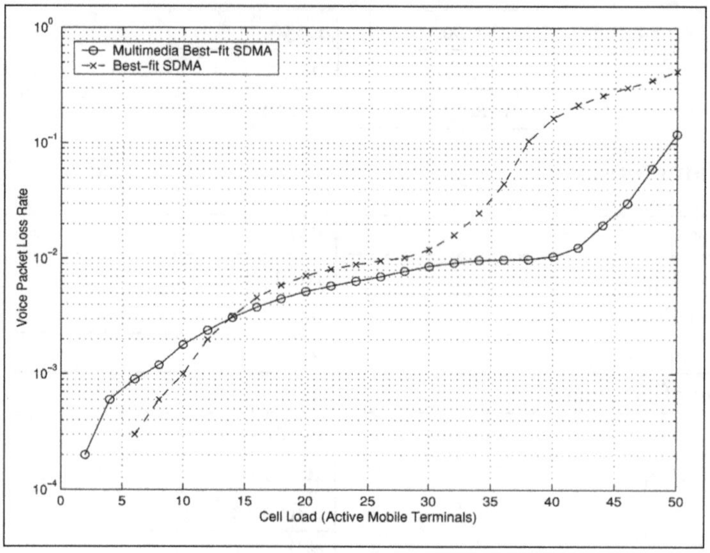


Figure: Packet loss rate of voice traffic.

Simulation results are provided to compare the throughput, delay, and packet loss of the improved protocol (multimedia best-fit SDMA) with that of best-fit SDMA. In the later case, the QoS requirements are worst case limited, e.g., the most strict SINR requirement of all traffic classes must be applied to all classes. In the case studied, the data traffic is assumed to have the highest SINR requirement at 12 dB, which is applied to both data and voice traffic in the simulation.

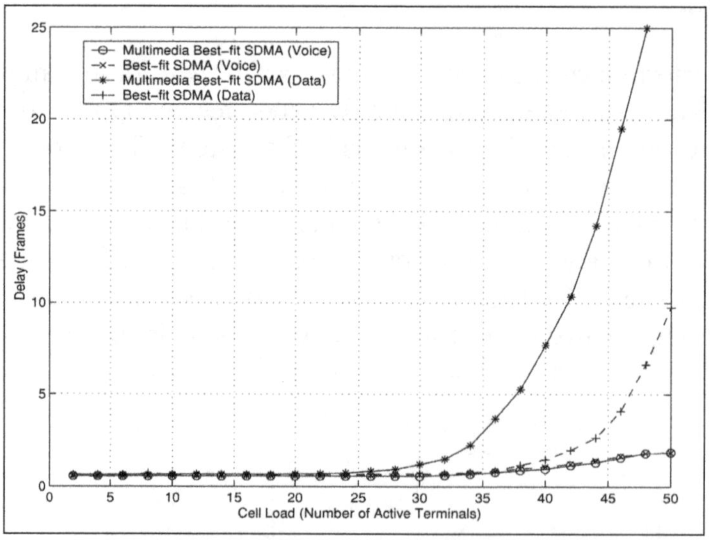


Figure: Delay characteristics of different traffics.

Figure compares the throughput characteristics of the two protocols. We observe about 30% of increase of throughput due to QoS-based prioritization. The is because that by taking the QoS into consideration, we are able to pack more voice packets into a frame with an acceptable packet loss penalty. Figure shows that the new protocol can keep the voice packet loss rate at an acceptable level even when there are a large number active terminals in the system. Assume that the voice packet loss must be kept below 1%, then the QoS-enhanced multimedia SDMA can accommodate up to 40 active terminals, which is 50% than the regular best-fit SDMA. This is because we cannot only pack more voice packets into a frame, but also we give higher priority to the voice packets based on their QoS requirements. The increased ability to handle voice traffic is paid by the increased delay of data traffic. Figure shows that for the multimedia SDMA protocol, the data traffic experiences longer delay. However, because the data traffic is relatively insensitive to delay, it will not cause severe performance degradation. This demonstrates that the new protocol is able to recognize the delay characteristics of different traffic and favors the more delay constrained traffic like voice traffic.

# 4

# Mobile Network and Transport Layer

A mobile network is a radio network with wireless last links distributed over land areas and served by cell sites or base stations. The transport layer is a conceptual division of methods in the layered architecture of protocols in the network stack in the OSI model and the internet protocol suite. Mobile internet protocol and transport control protocol are the protocols used in mobile networks. This chapter has been carefully written to provide an easy understanding of mobile networks and transport layers.

## Mobile Internet Protocol

Mobile IP (Internet Protocol) enables the transfer of information to and from mobile computers, such as laptops and wireless communications. The mobile computer can change its location to a foreign network and still access and communicate with and through the mobile computer's home network. The Solaris implementation of Mobile IP supports only IPv4.

Current versions of the Internet Protocol (IP) assume that the point at which a computer attaches to the Internet or a network is fixed and its IP address identifies the network to which it is attached. Datagrams are sent to a computer based on the location information contained in the IP address.

If a mobile computer, or mobile node, moves to a new network while keeping its IP address unchanged, its address does not reflect the new point of attachment. Consequently, existing routing protocols cannot route datagrams to the mobile node correctly. In this situation, you must reconfigure the mobile node with a different IP address representative of its new location, which is a cumbersome process. Thus, under the current Internet Protocol, if the mobile node moves without changing its address, it loses routing; but if it does change its address, it loses connections.

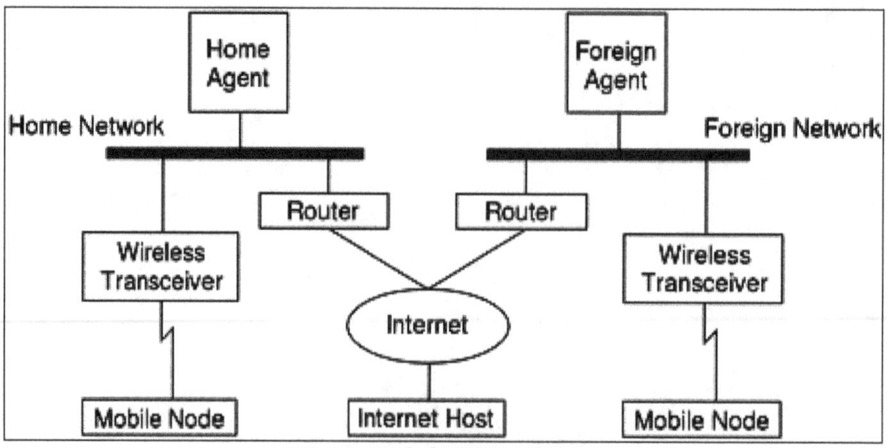


Figure: Mobile IP Topology.

Mobile IP solves this problem by allowing the mobile node to use two IP addresses: a fixed home address and a care-of address that changes at each new point of attachment. Mobile IP enables a computer to roam freely on the Internet or an organization's network while still maintaining the same home address. Consequently, computing activities are not disrupted when the user changes the computer's point of attachment to the Internet or an organization's network. Instead, the network is updated with the new location of the mobile node.

Using the previous illustration's Mobile IP topology, the following scenario shows how a datagram moves from one point to another within the Mobile IP framework:

- The Internet host sends a datagram to the mobile node using the mobile node's home address (normal IP routing process).
- If the mobile node is on its home network, the datagram is delivered through the normal IP process to the mobile node. Otherwise, the home agent picks up the datagram.
- If the mobile node is on a foreign network, the home agent forwards the datagram to the foreign agent.
- The foreign agent delivers the datagram to the mobile node.
- Datagrams from the mobile node to the Internet host are sent using normal IP routing procedures. If the mobile node is on a foreign network, the packets are delivered to the foreign agent. The foreign agent forwards the datagram to the Internet host.

In the case of wireless communications, the illustrations depict the use of wireless transceivers to transmit the datagrams to the mobile node. Also, all datagrams between the Internet host and the mobile node use the mobile node's home address regardless of whether the mobile node is on a home or foreign network. The care-of address is used only for communication with mobility agents and is never seen by the Internet host.

## Mobile IP Functional Entities

Mobile IP introduces the following new functional entities:

## How Mobile IP Works

Mobile IP enables routing of IP datagrams to mobile nodes. The mobile node's home address always identifies the mobile node, regardless of its current point of attachment to the Internet or an organization's network. When away from home, a care-of address associates the mobile node with its home address by providing information about the mobile node's current point of attachment to the Internet or an organization's network. Mobile IP uses a registration mechanism to register the care-of address with a home agent.

The home agent redirects datagrams from the home network to the care-of address by constructing a new IP header that contains the mobile node's care-of address as the destination IP address. This new header then encapsulates the original IP datagram, causing the mobile node's home address to have no effect on the encapsulated datagram's routing until it arrives at the care-of address. This type of encapsulation is also called tunneling. After arriving at the care-of address, each datagram is de-encapsulated and then delivered to the mobile node.

The following illustration shows a mobile node residing on its home network, Network A, before the mobile node moves to a foreign network, Network B. Both networks support Mobile IP. The mobile node is always associated with its home network by its permanent IP address, 128.226.3.30. Though Network A has a home agent, datagrams destined for the mobile node are delivered through the normal IP process.

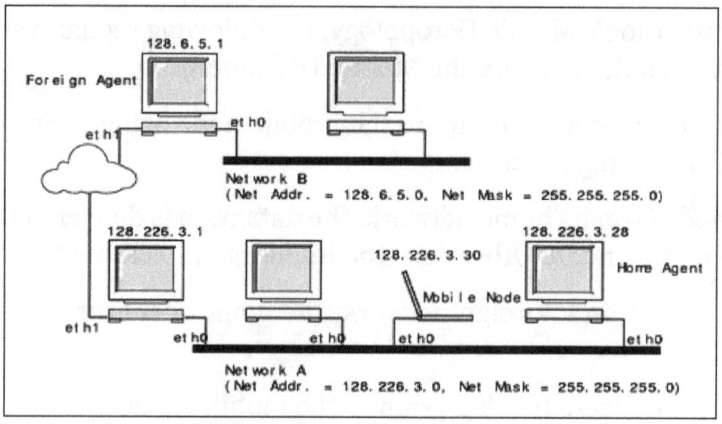


Figure: Mobile Node Residing on Home Network.

The following illustration shows the mobile node moving to a foreign network, Network B. Datagrams destined for the mobile node are intercepted by the home agent on the home network, Network A, encapsulated, and sent to the foreign agent on Network B. Upon receiving the encapsulated datagram, the foreign agent strips off the outer header and delivers the datagram to the mobile node visiting Network B.

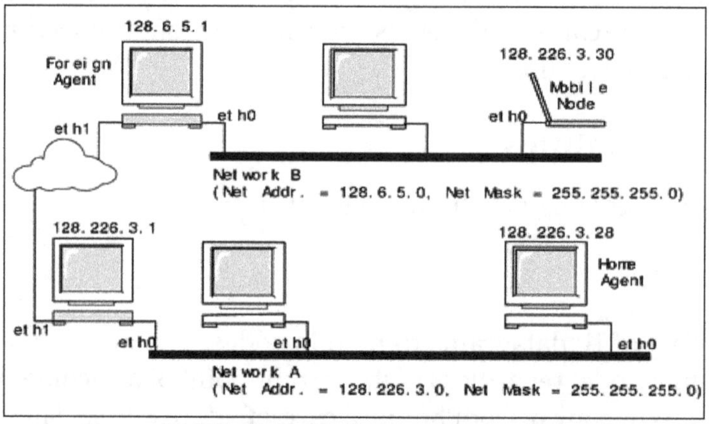


Figure: Mobile Node Moving to a Foreign Network.

The care-of address might belong to a foreign agent, or might be acquired by the mobile node through Dynamic Host Configuration Protocol (DHCP) or Point-to-Point Protocol (PPP). In the latter case, a mobile node is said to have a co-located care-of address.

The mobile node uses a special registration process to keep its home agent informed about its current location. Whenever a mobile node moves from its home network to a foreign network, or from one foreign network to another, it chooses a foreign agent on the new network and uses it to forward a registration message to its home agent.

Mobility agents (home agents and foreign agents) advertise their presence using agent advertisement messages. A mobile node can optionally solicit an agent advertisement message from any locally attached mobility agents through an agent solicitation message. A mobile node receives these agent advertisements and determines whether they are on its home network or a foreign network.

When the mobile node detects that it is located on its home network, it operates without mobility services. If returning to its home network from being registered elsewhere, the mobile node deregisters with its home agent.

## Mobile IP with Reverse Tunneling

The previous description of Mobile IP assumes that the routing within the Internet is independent of the data packet's source address. However, intermediate routers might check for a topologically correct source address. If an intermediate router does check, you should set up a reverse tunnel. By setting up a reverse tunnel from the mobile node's care-of address to the home agent, you ensure a topologically correct source address for the IP data packet. A mobile node can request a reverse tunnel between its foreign agent and its home agent when the mobile node registers. A reverse tunnel is a tunnel that starts at the mobile node's care-of address and terminates at the home agent. The following illustration shows the Mobile IP topology that uses a reverse tunnel.

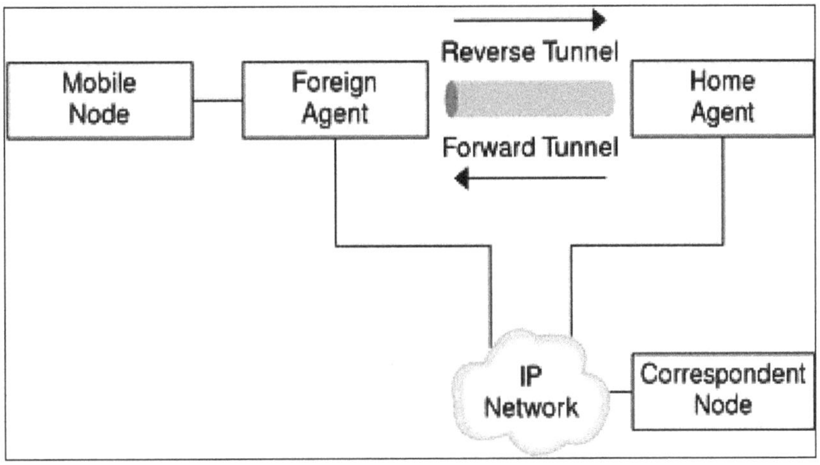


Figure: Mobile IP With a Reverse Tunnel.

## Limited Private Addresses Support

Mobile nodes that have private addresses which are not globally routable through the Internet require reverse tunnels. Solaris Mobile IP supports only privately addressed mobile nodes.

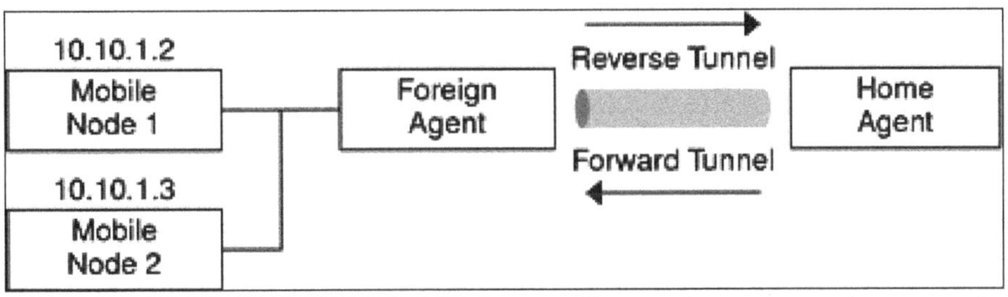


Figure: Privately Addressed Mobile Nodes Residing on the Same Foreign Network.

Enterprises employ private addresses when external connectivity is not required. Private addresses are not routable through the Internet. When a mobile node has a private address, the mobile node can only communicate with a correspondent node through a reverse tunnel. The privately addressed correspondent node must belong to the same home agent's administrative domain. The following illustration shows a network topology with two privately addressed mobile nodes that use the same care-of address when registered to the same foreign agent.

Because both privately addressed mobile nodes belong to the same administrative domain, the home agent knows how to route data packets between the two mobile nodes. Also, the foreign agent's care-of address and the home agent's IP address must be globally routable addresses.

It is possible to have two privately addressed mobile nodes with the same IP address residing on the same foreign network. This situation is only possible when each mobile node has a different home agent. Also, this situation is only possible when each mobile node is on different advertising subnets of a single foreign agent. The following illustration shows a network topology that depicts this case.

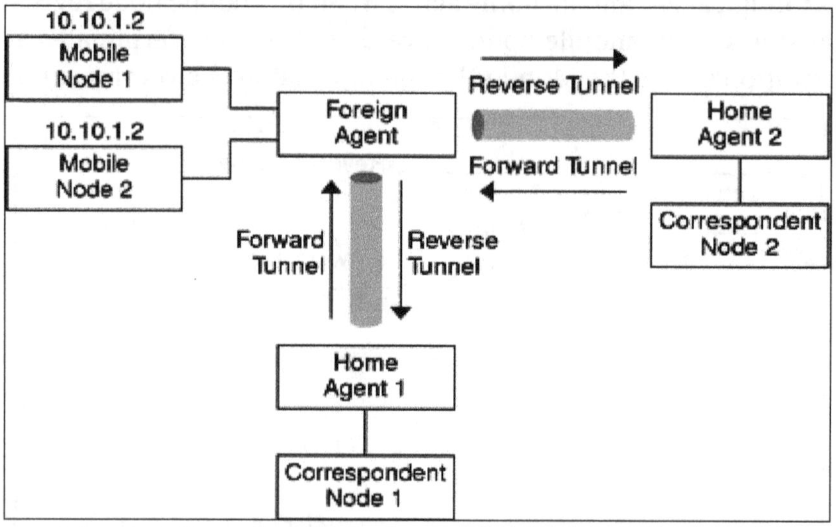


Figure: Privately Addressed Mobile Nodes Residing on Different Foreign Networks.

Because both privately addressed mobile nodes have the same IP address and because these mobile nodes belong to different home agent domains, the two nodes cannot communicate with each other. However, each node can communicate with nodes in its corresponding home agent's administrative domain through the reverse tunnel. For example, Mobile Node 2 can communicate with Correspondent Node 2 in the previous illustration.

## Care-of Addresses

Mobile IP provides the following alternative modes for the acquisition of a care-of address:

- A foreign agent provides a foreign agent care-of address through its agent advertisement messages. In this case, the care-of address is an IP address of the foreign agent. The foreign agent is the endpoint of the tunnel and, on receiving tunneled datagrams, de-encapsulates them and delivers the inner datagram to the mobile node. In this mode, many mobile nodes can share the same care-of address. This sharing reduces demands on the IPv4 address space and can also save bandwidth, because the forwarded packets, from the foreign agent to the mobile node, are not encapsulated. Saving bandwidth is important on wireless links.

- A mobile node acquires a co-located care-of address as a local IP address through some external means, which the mobile node then associates with one of its own network interfaces. The address might be dynamically acquired as a temporary address by the mobile node, such as through DHCP. The address might also be owned by the mobile node as a long-term address for its use only while visiting some foreign network. When using a co-located care-of address, the mobile node serves as the endpoint of the tunnel and performs de-encapsulation of the datagrams tunneled to it.

Co-located care-of address enables a mobile node to function without a foreign agent, for example, in networks that have not yet deployed a foreign agent.

If a mobile node is using a co-located care-of address, the mobile node must be located on the link identified by the network prefix of this care-of address. Otherwise, datagrams destined to the care-of address are undeliverable.

### Agent Discovery

A mobile node uses a method known as agent discovery to determine the following information:

- When the node has moved from one network to another.
- Whether the network is the node's home or a foreign network.
- What is the foreign agent care-of address offered by each foreign agent on that network.

Mobility agents transmit agent advertisements to advertise their services on a network. In the absence of agent advertisements, a mobile node can solicit advertisements. This is known as agent solicitation.

### Agent Advertisement

Mobile nodes use agent advertisements to determine their current point of attachment to the Internet or to an organization's network. An agent advertisement is an Internet Control Message Protocol (ICMP) router advertisement that has been extended to also carry a mobility agent advertisement extension.

A foreign agent can be too busy to serve additional mobile nodes. However, a foreign agent must continue to send agent advertisements. This way, mobile nodes that are already registered with it will know that they have not moved out of range of the foreign agent and that the foreign agent has not failed. Also, a foreign agent that supports reverse tunnels must send it's advertisements with the reverse tunnel flag set on.

### Agent Solicitation

Every mobile node should implement agent solicitation. The mobile node uses the same procedures, defaults, and constants for agent solicitation, as specified for ICMP router solicitation messages.

The rate at which a mobile node sends solicitations is limited by the mobile node. The mobile node can send three initial solicitations at a maximum rate of one per second while searching for an agent. After registering with an agent, the rate at which solicitations are sent is reduced, to limit the overhead on the local network.

## Mobile IP Registration

When the mobile node receives an agent advertisement, the mobile node registers through the foreign agent, even when the mobile node might be able to acquire its own co-located care-of address. This feature enables sites to restrict access to mobility services. Through agent advertisements, mobile nodes detect when they have moved from one subnet to another.

Mobile IP registration provides a flexible mechanism for mobile nodes to communicate their current reachability information to their home agent. The registration process enables mobile nodes to perform the following tasks:

- Request forwarding services when visiting a foreign network.
- Inform their home agent of their current care-of address.
- Renew a registration that is due to expire.
- Deregister when they return home.
- Request a reverse tunnel.

Registration messages exchange information between a mobile node, a foreign agent, and the home agent. Registration creates or modifies a mobility binding at the home agent, associating the mobile node's home address with its care-of address for the specified lifetime.

The registration process also enables mobile nodes to:

- Register with multiple foreign agents.
- Deregister specific care-of addresses while retaining other mobility bindings.
- Discover the address of a home agent if the mobile node is not configured with this information.

Mobile IP defines the following registration processes for a mobile node:

- If a mobile node is registering a foreign agent care-of address, the mobile node registers using that foreign agent.
- If a mobile node is using a co-located care-of address, and receives an agent advertisement from a foreign agent on the link on which it is using this care-of address, the mobile node registers using that foreign agent (or another foreign agent on this link).
- If a mobile node uses a co-located care-of address, the mobile node registers directly with its home agent.
- If a mobile node returns to its home network, the mobile node deregisters with its home agent.

These registration processes involve the exchange of registration requests and registration reply messages. When registering using a foreign agent, the registration process takes the following steps, which the subsequent illustration depicts:

- The mobile node sends a registration request to the prospective foreign agent to begin the registration process.
- The foreign agent processes the registration request and then relays it to the home agent.
- The home agent sends a registration reply to the foreign agent to grant or deny the request.
- The foreign agent processes the registration reply and then relays it to the mobile node to inform it of the disposition of its request.

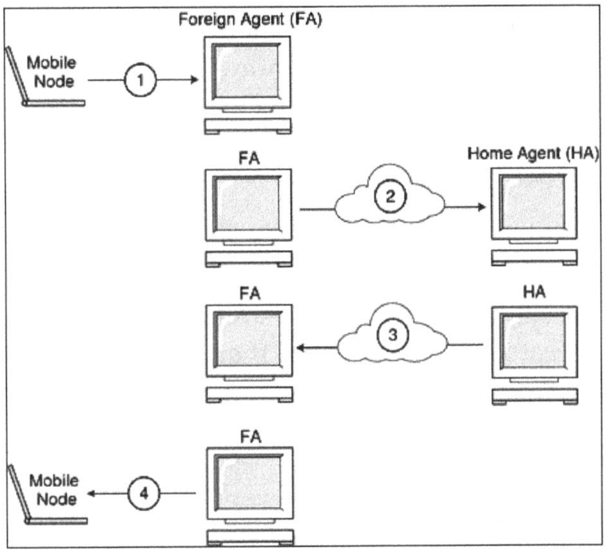


Figure: Mobile IP Registration Process.

When the mobile node registers directly with its home agent, the registration process requires only the following steps:

- The mobile node sends a deregistration request to the home agent.
- The home agent sends a registration reply to the mobile node, granting or denying the request.

Also, a reverse tunnel might be required by either the foreign agent or the home agent. If the foreign agent supports reverse tunneling, the mobile node uses the registration process to request a reverse tunnel. The mobile node does this by setting the reverse tunnel flag on in the mobile node's registration request.

## Network Access Identifier (NAI)

AAA servers, in use within the Internet, provide authentication and authorization services for dial-up computers. These services are likely to be equally valuable for mobile nodes using Mobile IP when the nodes are attempting to connect to foreign domains with AAA servers. AAA servers identify clients by using the Network Access Identifier (NAI). A mobile node can identify itself by including the NAI in the Mobile IP registration request.

Since the NAI is typically used to identify the mobile node uniquely, the mobile node's home address is not always necessary to provide that function. Thus, it is possible for a mobile node to authenticate itself, and be authorized for connection to the foreign domain, without even having a home address. To request that a home address be assigned, a message containing the mobile node NAI extension can set the home address field to zero in the registration request.

### Mobile IP Message Authentication

Each mobile node, foreign agent, and home agent supports a mobility security association between the various Mobile IP components, indexed by their security parameter index (SPI) and IP address. In the case of the mobile node, this address is its home address. Registration messages between a mobile node and its home agent are authenticated with the Mobile-home authentication extension. In addition to Mobile-home authentication, which is mandatory, you can use the optional Mobile-foreign agent and Home-foreign agent authentications.

### Mobile Node Registration Request

A mobile node registers with its home agent using a registration request message so that its home agent can create or modify a mobility binding for that mobile node (for example, with a new lifetime). The foreign agent can relay the registration request to the home agent. However, if the mobile node is registering a co-located care-of address, then the mobile node can send the registration request directly to the home agent.

### Registration Reply Message

A mobility agent returns a registration reply message to a mobile node that has sent a registration request message. If the mobile node is requesting service from a foreign agent, that foreign agent receives the reply from the home agent and subsequently relays it to the mobile node. The reply message contains the necessary codes to inform the mobile node about the status of its request, along with the lifetime granted by the home agent, which can be smaller than the original request. The registration reply can also contain a dynamic home address assignment.

### Foreign Agent Considerations

The foreign agent plays a mostly passive role in Mobile IP registration. A foreign agent adds all registered mobile nodes to its visitor table. It relays registration requests between mobile nodes and home agents, and, when it provides the care-of address, de-encapsulates datagrams for delivery to the mobile node. It also sends periodic agent advertisement messages to advertise its presence.

If reverse tunnels are supported, the foreign agent establishes appropriate routes to reverse tunnel all the data packets from the mobile node for a correspondent node. A foreign agent that supports reverse tunnels advertises that the reverse tunnel is supported for registration. Given the local policy, the foreign agent can deny a registration request when the reverse tunnel flag is not set. Also, the foreign agent can only distinguish two different mobile nodes with the same IP address when the mobile nodes visit on two different advertising interfaces.

## Home Agent Considerations

Home agents play an active role in the registration process. The home agent receives registration requests from the mobile node (perhaps relayed by a foreign agent), updates its record of the mobility bindings for this mobile node, and issues a suitable registration reply in response to each. The home agent also forwards packets to the mobile node when the mobile node is away from its home network.

A home agent might not have to have a physical subnet configured for mobile nodes. However, the home agent must recognize its mobile node's home address through the mipagent.conf file or some other mechanism when the home agent grants registration.

## Dynamic Home Agent Discovery

In some cases, the mobile node might not know its home agent address when the mobile node attempts to register. If the mobile node does not know its home agent address, the mobile node can use dynamic home agent address resolution to learn the address of its home agent. In this case, the mobile node sets the home agent field of the registration request to the subnet-directed broadcast address of the mobile node's home network. Each home agent that receives a registration request with a broadcast destination address rejects the mobile node's registration by returning a rejection registration reply. By doing so, the mobile node can use the home agent's unicast IP address indicated in the rejection reply when the mobile node next attempts registration.

## Routing Datagrams to and from Mobile Nodes

## Encapsulation Types

Home agents and foreign agents support tunneling datagrams using one of the available encapsulation methods (IP in IP Encapsulation, Minimal Encapsulation, or Generic Routing Encapsulation). Mobile nodes that use a co-located care-of address can receive tunneled datagrams using any encapsulation type.

## Unicast Datagram Routing

When registered on a foreign network, the mobile node chooses a default router using the following rules:

- If the mobile node is registered using a foreign agent care-of address, then the mobile node chooses its default router from among the router addresses advertised in the ICMP router advertisement portion of that agent advertisement message. The mobile node can also consider the IP source address of the agent advertisement as another possible choice for the IP address of a default router.
- If the mobile node is registered directly with its home agent using a co-located care-of address, then the mobile node chooses its default router from among those advertised in any ICMP router advertisement message that it receives. The chosen default router network prefix must match the mobile nodes externally obtained care-of address. If the mobile node's externally obtained care-of address matches the IP source address of the agent

advertisement under the network prefix, the mobile node can also consider that IP source address as another possible choice for the IP address of a default router.

- If the mobile node is registered, a foreign agent that supports reverse tunnels routes unicast datagrams from the mobile node to the home agent through the reverse tunnel.

### Broadcast Datagrams

When a home agent receives a broadcast datagram, it does not forward the datagram to any mobile nodes in its mobility binding list. However, the home agent does forward the datagram if a mobile node has requested forwarding of broadcast datagrams. For each registered mobile node, the home agent forwards received broadcast datagrams to the mobile node; the method depends on how the configuration of the home agent specifies categories of broadcast datagrams forwarded to mobile nodes. Broadcast datagrams over reverse tunnels are not supported.

### Multicast Datagram Routing

To receive multicasts, a mobile node joins the multicast group in one of the following ways:

- If a multicast router exists on the visited subnet, the mobile node uses this local multicast router. If the mobile node is using a co-located care-of address, it uses this address as the source IP address of its Internet Group Management Protocol (IGMP) messages. Otherwise, it uses its home address.
- If the mobile node's home agent is a multicast router, the mobile node can join groups using a bidirectional tunnel to its home agent. The mobile node tunnels IGMP messages to its home agent. The home agent then forwards multicast datagrams down the tunnel to the mobile node.

A mobile node that sends datagrams to a multicast group also has the following options:

- Send directly on the visited network.
- Send through a tunnel to its home agent.

Multicast routing depends on the IP source address. Therefore, a mobile node that sends multicast datagrams directly on the visited network uses a co-located care-of address as the IP source address. Similarly, a mobile node that tunnels a multicast datagram to its home agent uses its home address as the IP source address of both the multicast datagram and the encapsulating datagram. This second option assumes that the home agent is a multicast router. In the case of reverse tunnels, multicast datagrams are not routed through reverse tunnels. The multicast datagrams are routed as previously described.

### Security Considerations

In many cases, mobile computers use wireless links to connect to the network. Wireless links are particularly vulnerable to passive eavesdropping, active replay attacks, and other active attacks.

Though Mobile IP cannot reduce or eliminate this vulnerability, Mobile IP can authenticate the Mobile IP messages. The default algorithm used is MD5, with a key size of 128 bits. The default operational mode requires that this 128–bit key precede and succeed the data to be hashed. The foreign agent also supports authentication using MD5 and key sizes of 128 bits or greater, with manual key distribution. Mobile IP can support more authentication algorithms, algorithm modes, key distribution methods, and key sizes.

Tunneling can be a significant vulnerability, especially if registration is not authenticated. Also, the Address Resolution Protocol (ARP) is not authenticated, and can potentially be used to steal another host's traffic.

# Transmission Control Protocol

The Transmission Control Protocol (TCP) is intended for use as a highly reliable host-to-host protocol between hosts in packet-switched computer communication networks, and in interconnected systems of such networks.

## Motivation

Computer communication systems are playing an increasingly important role in military, government, and civilian environments. As strategic and tactical computer communication networks are developed and deployed, it is essential to provide means of interconnecting them and to provide standard interprocess communication protocols which can support a broad range of applications. In anticipation of the need for such standards, the Deputy Undersecretary of Defense for Research and Engineering has declared the Transmission Control Protocol (TCP) described herein to be a basis for DoD-wide inter-process communication protocol standardization.

TCP is a connection-oriented, end-to-end reliable protocol designed to fit into a layered hierarchy of protocols which support multi-network applications. The TCP provides for reliable inter-process communication between pairs of processes in host computers attached to distinct but interconnected computer communication networks. Very few assumptions are made as to the reliability of the communication protocols below the TCP layer. TCP assumes it can obtain a simple, potentially unreliable datagram service from the lower level protocols. In principle, the TCP should be able to operate above a wide spectrum of communication systems ranging from hard-wired connections to packet-switched or circuit-switched networks.

TCP is based on concepts first described by Cerf and Kahn in. The TCP fits into a layered protocol architecture just above a basic Internet Protocol which provides a way for the TCP to send and receive variable-length segments of information enclosed in internet datagram "envelopes". The internet datagram provides a means for addressing source and destination TCPs in different networks. The internet protocol also deals with any fragmentation or reassembly of the TCP segments required to achieve transport and delivery through multiple networks and interconnecting gateways. The internet protocol also carries information on the precedence, security classification and compartmentation of the TCP segments, so this information can be communicated end-to-end across multiple networks.

Table: Protocol Layering.

| Higher-level |
|---|
| TCP |
| Internet protocol |
| Communication network |

Some computer systems will be connected to networks via front-end computers which house the TCP and internet protocol layers, as well as network specific software. The TCP specification describes an interface to the higher level protocols which appears to be implementable even for the front-end case, as long as a suitable host-to-front end protocol is implemented.

## Scope

The TCP is intended to provide a reliable process-to-process communication service in a multinetwork environment. The TCP is intended to be a host-to-host protocol in common use in multiple networks.

## Interfaces

The TCP interfaces on one side to user or application processes and on the other side to a lower level protocol such as Internet Protocol.

The interface between an application process and the TCP is illustrated in reasonable detail. This interface consists of a set of calls much like the calls an operating system provides to an application process for manipulating files. For example, there are calls to open and close connections and to send and receive data on established connections. It is also expected that the TCP can asynchronously communicate with application programs. Although considerable freedom is permitted to TCP implementors to design interfaces which are appropriate to a particular operating system environment, a minimum functionality is required at the TCP/user interface for any valid implementation.

The interface between TCP and lower level protocol is essentially unspecified except that it is assumed there is a mechanism whereby the two levels can asynchronously pass information to each other. Typically, one expects the lower level protocol to specify this interface. TCP is designed to work in a very general environment of interconnected networks.

## Operation

The primary purpose of the TCP is to provide reliable, securable logical circuit or connection service between pairs of processes. To provide this service on top of a less reliable internet communication system requires facilities in the following areas:

- Basic Data Transfer.
- Reliability.
- Flow Control.
- Multiplexing.

- Connections.
- Precedence and Security.

The basic operation of the TCP in each of these areas is described in the following paragraphs:

### Basic Data Transfer

The TCP is able to transfer a continuous stream of octets in each direction between its users by packaging some number of octets into segments for transmission through the internet system. In general, the TCPs decide when to block and forward data at their own convenience. Sometimes users need to be sure that all the data they have submitted to the TCP has been transmitted. For this purpose a push function is defined. To assure that data submitted to a TCP is actually transmitted the sending user indicates that it should be pushed through to the receiving user. A push causes the TCPs to promptly forward and deliver data up to that point to the receiver. The exact push point might not be visible to the receiving user and the push function does not supply a record boundary marker.

### Reliability

The TCP must recover from data that is damaged, lost, duplicated, or delivered out of order by the internet communication system. This is achieved by assigning a sequence number to each octet transmitted, and requiring a positive acknowledgment (ACK) from the receiving TCP. If the ACK is not received within a timeout interval, the data is retransmitted. At the receiver, the sequence numbers are used to correctly order segments that may be received out of order and to eliminate duplicates. Damage is handled by adding a checksum to each segment transmitted, checking it at the receiver, and discarding damaged segments. As long as the TCPs continue to function properly and the internet system does not become completely partitioned, no transmission errors will affect the correct delivery of data. TCP recovers from internet communication system errors.

### Flow Control

TCP provides a means for the receiver to govern the amount of data sent by the sender. This is achieved by returning a "window" with every ACK indicating a range of acceptable sequence numbers beyond the last segment successfully received. The window indicates an allowed number of octets that the sender may transmit before receiving further permission.

### Multiplexing

To allow for many processes within a single Host to use TCP communication facilities simultaneously, the TCP provides a set of addresses or ports within each host. Concatenated with the network and host addresses from the internet communication layer, this forms a socket. A pair of sockets uniquely identifies each connection. That is, a socket may be simultaneously used in multiple connections.

The binding of ports to processes is handled independently by each Host. However, it proves useful to attach frequently used processes (e.g., a "logger" or timesharing service) to fixed sockets which are made known to the public. These services can then be accessed through the known addresses. Establishing and learning the port addresses of other processes may involve more dynamic mechanisms.

## Connections

The reliability and flow control mechanisms described above require that TCPs initialize and maintain certain status information for each data stream. The combination of this information, including sockets, sequence numbers, and window sizes, is called a connection. Each connection is uniquely specified by a pair of sockets identifying its two sides.

When two processes wish to communicate, their TCP's must first establish a connection (initialize the status information on each side). When their communication is complete, the connection is terminated or closed to free the resources for other uses.

Since connections must be established between unreliable hosts and over the unreliable internet communication system, a handshake mechanism with clock-based sequence numbers is used to avoid erroneous initialization of connections.

## Precedence and Security

The users of TCP may indicate the security and precedence of their communication. Provision is made for default values to be used when these features are not needed.

## Functional Specification

## Header Format

TCP segments are sent as internet datagrams. The Internet Protocol header carries several information fields, including the source and destination host addresses. A TCP header follows the internet header, supplying information specific to the TCP protocol. This division allows for the existence of host level protocols other than TCP.

```
 0                   1                   2                   3
 0 1 2 3 4 5 6 7 8 9 0 1 2 3 4 5 6 7 8 9 0 1 2 3 4 5 6 7 8 9 0 1
+-+-+-+-+-+-+-+-+-+-+-+-+-+-+-+-+-+-+-+-+-+-+-+-+-+-+-+-+-+-+-+-+
|          Source Port          |       Destination Port        |
+-+-+-+-+-+-+-+-+-+-+-+-+-+-+-+-+-+-+-+-+-+-+-+-+-+-+-+-+-+-+-+-+
|                        Sequence Number                        |
+-+-+-+-+-+-+-+-+-+-+-+-+-+-+-+-+-+-+-+-+-+-+-+-+-+-+-+-+-+-+-+-+
|                    Acknowledgment Number                      |
+-+-+-+-+-+-+-+-+-+-+-+-+-+-+-+-+-+-+-+-+-+-+-+-+-+-+-+-+-+-+-+-+
|  Data |           |U|A|P|R|S|F|                               |
| Offset| Reserved  |R|C|S|S|Y|I|            Window             |
|       |           |G|K|H|T|N|N|                               |
+-+-+-+-+-+-+-+-+-+-+-+-+-+-+-+-+-+-+-+-+-+-+-+-+-+-+-+-+-+-+-+-+
|           Checksum            |         Urgent Pointer        |
+-+-+-+-+-+-+-+-+-+-+-+-+-+-+-+-+-+-+-+-+-+-+-+-+-+-+-+-+-+-+-+-+
|                    Options                    |    Padding    |
+-+-+-+-+-+-+-+-+-+-+-+-+-+-+-+-+-+-+-+-+-+-+-+-+-+-+-+-+-+-+-+-+
|                             data                              |
+-+-+-+-+-+-+-+-+-+-+-+-+-+-+-+-+-+-+-+-+-+-+-+-+-+-+-+-+-+-+-+-+
```

TCP Header Format.

Note that one tick mark represents one bit position.

Source Port: 16 bits:

- The source port number.

Destination Port: 16 bits:

- The destination port number.

Sequence Number: 32 bits:

- The sequence number of the first data octet in this segment (except when SYN is present). If SYN is present the sequence number is the initial sequence number (ISN) and the first data octet is ISN+1.

Acknowledgment Number: 32 bits:

- If the ACK control bit is set this field contains the value of the next sequence number the sender of the segment is expecting to receive. Once a connection is established this is always sent.

Data Offset: 4 bits:

- The number of 32 bit words in the TCP Header. This indicates where the data begins. The TCP header (even one including options) is an integral number of 32 bits long.

Reserved: 6 bits:

- Reserved for future use. Must be zero.

Control Bits: 6 bits (from left to right):

- URG: Urgent Pointer field significant.
- ACK: Acknowledgment field significant.
- PSH: Push Function.
- RST: Reset the connection.
- SYN: Synchronize sequence numbers.
- FIN: No more data from sender.

Window: 16 bits:

- The number of data octets beginning with the one indicated in the acknowledgment field which the sender of this segment is willing to accept.

Checksum: 16 bits:

- The checksum field is the 16 bit one's complement of the one's complement sum of all 16 bit words in the header and text. If a segment contains an odd number of header and text octets to be checksummed, the last octet is padded on the right with zeros to form a 16 bit word for checksum purposes. The pad is not transmitted as part of the segment. While computing the checksum, the checksum field itself is replaced with zeros.

The checksum also covers a 96 bit pseudo header conceptually prefixed to the TCP header. This pseudo header contains the Source Address, the Destination Address, the Protocol, and TCP length. This gives the TCP protection against misrouted segments. This information is carried in the Internet Protocol and is transferred across the TCP/Network interface in the arguments or results of calls by the TCP on the IP.

```
+--------+--------+--------+--------+
|           Source Address          |
+--------+--------+--------+--------+
|         Destination Address       |
+--------+--------+--------+--------+
|  zero  |  PTCL  |    TCP Length   |
+--------+--------+--------+--------+
```

The TCP Length is the TCP header length plus the data length in octets (this is not an explicitly transmitted quantity, but is computed), and it does not count the 12 octets of the pseudo header.

Urgent Pointer: 16 bits:

- This field communicates the current value of the urgent pointer as a positive offset from the sequence number in this segment. The urgent pointer points to the sequence number of the octet following the urgent data. This field is only be interpreted in segments with the URG control bit set. Options: variable.

Options may occupy space at the end of the TCP header and are a multiple of 8 bits in length. All options are included in the checksum. An option may begin on any octet boundary. There are two cases for the format of an option:

- Case 1: A single octet of option-kind.
- Case 2: An octet of option-kind, an octet of option-length, and the actual option-data octets.

The option-length counts the two octets of option-kind and option-length as well as the option-data octets.

The list of options may be shorter than the data offset field might imply. The content of the header beyond the End-of-Option option must be header padding (i.e., zero).

Currently defined options include (kind indicated in octal):

| Kind | Length | Meaning |
|---|---|---|
| 0 | - | End of option list |

| | | |
|---|---|---|
| 1 | - | No-Operation. |
| 2 | 4 | Maximum Segment Size |

## Specific Option Definitions End of Option List

```
+--------+
|00000000|
+--------+
```

Kind=0:

- This option code indicates the end of the option list. This might not coincide with the end of the TCP header according to the Data Offset field. This is used at the end of all options, not the end of each option, and need only be used if the end of the options would not otherwise coincide with the end of the TCP header.

## No-Operation

```
+--------+
|00000001|
+--------+
```

Kind=1:

- This option code may be used between options, for example, to align the beginning of a subsequent option on a word boundary.
- There is no guarantee that senders will use this option, so receivers must be prepared to process options even if they do not begin on a word boundary.

## Maximum Segment Size

```
+--------+--------+---------+--------+
|00000010|00000100| max seg size |
+--------+--------+---------+--------+
```

Kind=2 Length=4:

Maximum Segment Size Option Data: 16 bits.

- If this option is present, then it communicates the maximum receive segment size at the TCP which sends this segment. This field must only be sent in the initial connection request (i.e., in segments with the SYN control bit set). If this option is not used, any segment size is allowed.

## Padding: Variable

The TCP header padding is used to ensure that the TCP header ends and data begins on a 32 bit boundary. The padding is composed of zeros.

## Terminology

Before we can discuss very much about the operation of the TCP we need to introduce some detailed terminology. The maintenance of a TCP connection requires the remembering of several variables. We conceive of these variables being stored in a connection record called a Transmission Control Block or TCB. Among the variables stored in the TCB are the local and remote socket numbers, the security and precedence of the connection, pointers to the user's send and receive buffers, pointers to the retransmit queue and to the current segment. In addition several variables relating to the send and receive sequence numbers are stored in the TCB.

Send Sequence Variables:

- SND.UNA - send unacknowledged.
- SND.NXT - send next.
- SND.WND - send window.
- SND.UP - send urgent pointer.
- SND.WL1 - segment sequence number used for last window update.
- SND.WL2 - segment acknowledgment number used for last window update.
- ISS - initial send sequence number.

Receive Sequence Variables:

- RCV.NXT - receive next.
- RCV.WND - receive window.
- RCV.UP - receive urgent pointer.
- IRS - initial receive sequence number.

The following diagrams may help to relate some of these variables to the sequence space.

Send Sequence Space:

```
      1          2          3          4
----------|----------|----------|----------
       SND.UNA    SND.NXT    SND.UNA
                            +SND.WND
```

1 - old sequence numbers which have been acknowledged.

2 - sequence numbers of unacknowledged data.

3 - sequence numbers allowed for new data transmission.

4 - future sequence numbers which are not yet allowed.

Send Sequence Space:

The send window is the portion of the sequence space labeled 3 in figure.

Receive Sequence Space:

```
        1              2             3
   --------------|--------------|--------------

            RCV.NXT       RCV.NXT

                          +RCV.WND
```

1 - old sequence numbers which have been acknowledged.

2 - sequence numbers allowed for new reception.

3 - future sequence numbers which are not yet allowed.

Receive Sequence Space:

- The receive window is the portion of the sequence space labeled 2 in figure. There are also some variables used frequently in the discussion that take their values from the fields of the current segment.

Current Segment Variables:

- SEG.SEQ - segment sequence number.
- SEG.ACK - segment acknowledgment number.
- SEG.LEN - segment length.
- SEG.WND - segment window.
- SEG.UP - segment urgent pointer.
- SEG.PRC - segment precedence value.

A connection progresses through a series of states during its lifetime. The states are: LISTEN, SYN-SENT, SYN-RECEIVED, ESTABLISHED, FIN-WAIT-1, FIN-WAIT-2, CLOSE-WAIT, CLOSING, LAST-ACK, TIME-WAIT, and the fictional state CLOSED. CLOSED is fictional because it represents the state when there is no TCB, and therefore, no connection. Briefly the meanings of the states are:

- LISTEN - represents waiting for a connection request from any remote TCP and port.
- SYN-SENT - represents waiting for a matching connection request after having sent a connection request.
- SYN-RECEIVED - represents waiting for a confirming connection request acknowledgment after having both received and sent a connection request.

- ESTABLISHED - represents an open connection, data received can be delivered to the user. The normal state for the data transfer phase of the connection.
- FIN-WAIT-1 - represents waiting for a connection termination request from the remote TCP, or an acknowledgment of the connection termination request previously sent.
- FIN-WAIT-2 - represents waiting for a connection termination request from the remote TCP.
- CLOSE-WAIT - represents waiting for a connection termination request from the local user.
- CLOSING - represents waiting for a connection termination request acknowledgment from the remote TCP.
- LAST-ACK - represents waiting for an acknowledgment of the connection termination request previously sent to the remote TCP (which includes an acknowledgment of its connection termination request).
- TIME-WAIT - represents waiting for enough time to pass to be sure the remote TCP received the acknowledgment of its connection termination request.
- CLOSED - represents no connection state at all.

A TCP connection progresses from one state to another in response to events. The events are the user calls, OPEN, SEND, RECEIVE, CLOSE, ABORT, and STATUS; the incoming segments, particularly those containing the SYN, ACK, RST and FIN flags; and timeouts.

The state diagram illustrates only state changes, together with the causing events and resulting actions, but addresses neither error conditions nor actions which are not connected with state changes.

## Sequence Numbers

A fundamental notion in the design is that every octet of data sent over a TCP connection has a sequence number. Since every octet is sequenced, each of them can be acknowledged. The acknowledgment mechanism employed is cumulative so that an acknowledgment of sequence number X indicates that all octets up to but not including X have been received. This mechanism allows for straight-forward duplicate detection in the presence of retransmission. Numbering of octets within a segment is that the first data octet immediately following the header is the lowest numbered, and the following octets are numbered consecutively.

It is essential to remember that the actual sequence number space is finite, though very large. This space ranges from 0 to 2**32 - 1. Since the space is finite, all arithmetic dealing with sequence numbers must be performed modulo 2**32. This unsigned arithmetic preserves the relationship of sequence numbers as they cycle from 2**32 - 1 to 0 again. There are some subtleties to computer modulo arithmetic, so great care should be taken in programming the comparison of such values. The symbol "=<" means "less than or equal" (modulo 2**32).

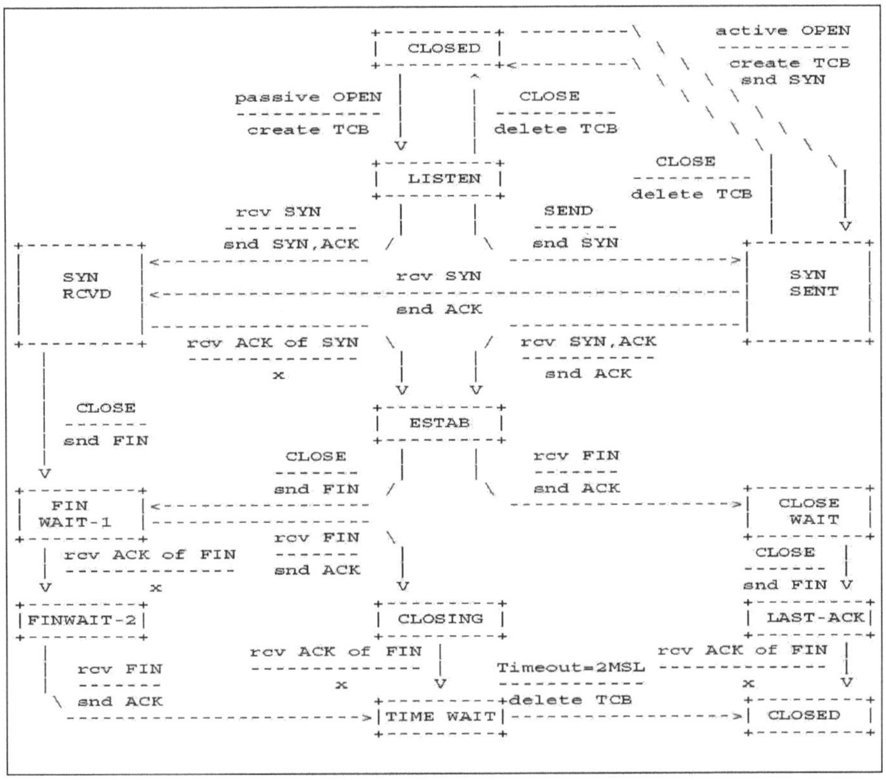


Figure: TCP Connection State Diagram.

The typical kinds of sequence number comparisons which the TCP must perform include:

- Determining that an acknowledgment refers to some sequence number sent but not yet acknowledged.
- Determining that all sequence numbers occupied by a segment have been acknowledged (e.g., to remove the segment from a retransmission queue).
- Determining that an incoming segment contains sequence numbers which are expected (i.e., that the segment "overlaps" the receive window).

In response to sending data the TCP will receive acknowledgments. The following comparisons are needed to process the acknowledgments.

- SND.UNA = oldest unacknowledged sequence number.
- SND.NXT = next sequence number to be sent.
- SEG.ACK = acknowledgment from the receiving TCP (next sequence number expected by the receiving TCP).
- SEG.SEQ = first sequence number of a segment.
- SEG.LEN = the number of octets occupied by the data in the segment (counting SYN and FIN).
- SEG.SEQ+SEG.LEN-1 = last sequence number of a segment.

A new acknowledgment (called an "acceptable ack"), is one for which the inequality below holds:

- SND.UNA < SEG.ACK =< SND.NXT

A segment on the retransmission queue is fully acknowledged if the sum of its sequence number and length is less or equal than the acknowledgment value in the incoming segment.

When data is received the following comparisons are needed:

RCV.NXT = next sequence number expected on an incoming segments, and is the left or lower edge of the receive window.

- RCV.NXT+RCV.WND-1 = last sequence number expected on an incoming segment, and is the right or upper edge of the receive window.
- SEG.SEQ = first sequence number occupied by the incoming segment.
- SEG.SEQ+SEG.LEN-1 = last sequence number occupied by the incoming segment.

A segment is judged to occupy a portion of valid receive sequence space if,

- RCV.NXT =< SEG.SEQ < RCV.NXT+RCV.WND

or

- RCV.NXT =< SEG.SEQ+SEG.LEN-1 < RCV.NXT+RCV.WND

The first part of this test checks to see if the beginning of the segment falls in the window, the second part of the test checks to see if the end of the segment falls in the window; if the segment passes either part of the test it contains data in the window.

Actually, it is a little more complicated than this. Due to zero windows and zero length segments, we have four cases for the acceptability of an incoming segment:

| Segment length | Receive Window | Test |
|---|---|---|
| 0 | 0 | SEG.SEQ = RCV.NXT |
| 0 | >0 | RCV.NXT =< SEG.SEQ < RCV.NXT+RCV.WND |
| >0 | 0 | not acceptable |
| >0 | >0 | RCV.NXT =< SEG.SEQ < RCV.NXT+RCV.WND or RCV.NXT =< SEG.SEQ+SEG.LEN-1 < RCV.NXT+RCV.WND |

Note that when the receive window is zero no segments should be acceptable except ACK segments. Thus, it is be possible for a TCP to maintain a zero receive window while transmitting data and receiving ACKs. However, even when the receive window is zero, a TCP must process the RST and URG fields of all incoming segments.

We have taken advantage of the numbering scheme to protect certain control information as well. This is achieved by implicitly including some control flags in the sequence space so they can be retransmitted and acknowledged without confusion (i.e., one and only one copy of the control will be acted upon). Control information is not physically carried in the segment data space. Consequently, we must adopt rules for implicitly assigning sequence numbers to control. The SYN and FIN are the only controls requiring this protection, and these controls are used only at connection opening and closing. For sequence number purposes, the SYN is considered to occur before the first actual data octet of the segment in which it occurs, while the FIN is considered to occur after the last actual data octet in a segment in which it occurs. The segment length (SEG.LEN) includes both data and sequence space occupying controls. When a SYN is present then SEG.SEQ is the sequence number of the SYN.

## Initial Sequence Number Selection

The protocol places no restriction on a particular connection being used over and over again. A connection is defined by a pair of sockets. New instances of a connection will be referred to as incarnations of the connection. The problem that arises from this is -- "how does the TCP identify duplicate segments from previous incarnations of the connection?" This problem becomes apparent if the connection is being opened and closed in quick succession, or if the connection breaks with loss of memory and is then re-established.

To avoid confusion we must prevent segments from one incarnation of a connection from being used while the same sequence numbers may still be present in the network from an earlier incarnation. We want to assure this, even if a TCP crashes and loses all knowledge of the sequence numbers it has been using. When new connections are created, an initial sequence number (ISN) generator is employed which selects a new 32 bit ISN. The generator is bound to a (possibly fictitious) 32 bit clock whose low order bit is incremented roughly every 4 microseconds. Thus, the ISN cycles approximately every 4.55 hours. Since we assume that segments will stay in the network no more than the Maximum Segment Lifetime (MSL) and that the MSL is less than 4.55 hours we can reasonably assume that ISN's will be unique.

For each connection there is a send sequence number and a receive sequence number. The initial send sequence number (ISS) is chosen by the data sending TCP, and the initial receive sequence number (IRS) is learned during the connection establishing procedure.

For a connection to be established or initialized, the two TCPs must synchronize on each other's initial sequence numbers. This is done in an exchange of connection establishing segments carrying a control bit called "SYN" (for synchronize) and the initial sequence numbers. As shorthand, segments carrying the SYN bit are also called "SYNs". Hence, the solution requires a suitable mechanism for picking an initial sequence number and a slightly involved handshake to exchange the ISN's.

The synchronization requires each side to send it's own initial sequence number and to receive a confirmation of it in acknowledgment from the other side. Each side must also receive the other side's initial sequence number and send a confirming acknowledgment.

- A --> B SYN my sequence number is X.
- A <-- B ACK your sequence number is X.
- A <-- B SYN my sequence number is Y.
- A --> B ACK your sequence number is Y.

Because steps 2 and 3 can be combined in a single message this is called the three way (or three message) handshake.

A three way handshake is necessary because sequence numbers are not tied to a global clock in the network, and TCPs may have different mechanisms for picking the ISN's. The receiver of the first SYN has no way of knowing whether the segment was an old delayed one or not, unless it remembers the last sequence number used on the connection (which is not always possible), and so it must ask the sender to verify this SYN.

## Knowing when to Keep Quiet

To be sure that a TCP does not create a segment that carries a sequence number which may be duplicated by an old segment remaining in the network, the TCP must keep quiet for a maximum segment lifetime (MSL) before assigning any sequence numbers upon starting up or recovering from a crash in which memory of sequence numbers in use was lost. For this specification the MSL is taken to be 2 minutes. This is an engineering choice, and may be changed if experience indicates it is desirable to do so. Note that if a TCP is reinitialized in some sense, yet retains its memory of sequence numbers in use, then it need not wait at all; it must only be sure to use sequence numbers larger than those recently used.

## The TCP Quiet Time Concept

This specification provides that hosts which "crash" without retaining any knowledge of the last sequence numbers transmitted on each active (i.e., not closed) connection shall delay emitting any TCP segments for at least the agreed Maximum Segment Lifetime (MSL) in the internet system of which the host is a part. In the paragraphs below, an explanation for this specification is given. TCP implementors may violate the "quiet time" restriction, but only at the risk of causing some old data to be accepted as new or new data rejected as old duplicated by some receivers in the internet system.

TCPs consume sequence number space each time a segment is formed and entered into the network output queue at a source host. The duplicate detection and sequencing algorithm in the TCP protocol relies on the unique binding of segment data to sequence space to the extent that sequence numbers will not cycle through all 2**32 values before the segment data bound to those sequence numbers has been delivered and acknowledged by the receiver and all duplicate copies of the segments have "drained" from the internet. Without such an assumption, two distinct TCP segments could conceivably be assigned the same or overlapping sequence numbers, causing confusion at the receiver as to which data is new and which is old. Remember that each segment is bound to as many consecutive sequence numbers as there are octets of data in the segment.

Under normal conditions, TCPs keep track of the next sequence number to emit and the oldest awaiting acknowledgment so as to avoid mistakenly using a sequence number over before its first use has been acknowledged. This alone does not guarantee that old duplicate data is drained from the net, so the sequence space has been made very large to reduce the probability that a wandering duplicate will cause trouble upon arrival. At 2 megabits/sec. it takes 4.5 hours to use up 2**32 octets of sequence space. Since the maximum segment lifetime in the net is not likely to exceed a few tens of seconds, this is deemed ample protection for foreseeable nets, even if data rates escalate to lo's of megabits/sec. At 100 megabits/sec, the cycle time is 5.4 minutes which may be a little short, but still within reason.

The basic duplicate detection and sequencing algorithm in TCP can be defeated, however, if a source TCP does not have any memory of the sequence numbers it last used on a given connection. For example, if the TCP were to start all connections with sequence number 0, then upon crashing and restarting, a TCP might re-form an earlier connection (possibly after half-open connection resolution) and emit packets with sequence numbers identical to or overlapping with packets still in the network which were emitted on an earlier incarnation of the same connection. In the absence of knowledge about the sequence numbers used on a particular connection, the TCP specification recommends that the source delay for MSL seconds before emitting segments on the connection, to allow time for segments from the earlier connection incarnation to drain from the system.

Even hosts which can remember the time of day and used it to select initial sequence number values are not immune from this problem (i.e., even if time of day is used to select an initial sequence number for each new connection incarnation).

Suppose, for example, that a connection is opened starting with sequence number S. Suppose that this connection is not used much and that eventually the initial sequence number function (ISN(t)) takes on a value equal to the sequence number, say S1, of the last segment sent by this TCP on a particular connection. Now suppose, at this instant, the host crashes, recovers, and establishes a new incarnation of the connection. The initial sequence number chosen is S1 = ISN(t) -- last used sequence number on old incarnation of connection! If the recovery occurs quickly enough, any old duplicates in the net bearing sequence numbers in the neighborhood of S1 may arrive and be treated as new packets by the receiver of the new incarnation of the connection. The problem is that the recovering host may not know for how long it crashed nor does it know whether there are still old duplicates in the system from earlier connection incarnations.

One way to deal with this problem is to deliberately delay emitting segments for one MSL after recovery from a crash- this is the "quite time" specification. Hosts which prefer to avoid waiting are willing to risk possible confusion of old and new packets at a given destination may choose not to wait for the "quite time". Implementors may provide TCP users with the ability to select on a connection by connection basis whether to wait after a crash, or may informally implement the "quite time" for all connections. Obviously, even where a user selects to "wait," this is not necessary after the host has been "up" for at least MSL seconds.

To summarize: every segment emitted occupies one or more sequence numbers in the sequence space, the numbers occupied by a segment are "busy" or "in use" until MSL seconds have passed, upon crashing a block of space-time is occupied by the octets of the last emitted segment, if a new connection is started too soon and uses any of the sequence numbers in the space-time footprint of the last segment of the previous connection incarnation, there is a potential sequence number overlap area which could cause confusion at the receiver.

## Establishing a Connection

The "three-way handshake" is the procedure used to establish a connection. This procedure normally is initiated by one TCP and responded to by another TCP. The procedure also works if two TCP simultaneously initiate the procedure. When simultaneous attempt occurs, each TCP receives a "SYN" segment which carries no acknowledgment after it has sent a "SYN". Of course, the arrival of an old duplicate "SYN" segment can potentially make it appear, to the recipient, that a simultaneous connection initiation is in progress. Proper use of "reset" segments can disambiguate these cases.

Several examples of connection initiation follow. Although these examples do not show connection synchronization using data-carrying segments, this is perfectly legitimate, so long as the receiving TCP doesn't deliver the data to the user until it is clear the data is valid (i.e., the data must be buffered at the receiver until the connection reaches the ESTABLISHED state). The three-way handshake reduces the possibility of false connections. It is the implementation of a trade-off between memory and messages to provide information for this checking.

The simplest three-way handshake is shown in table below. The tables should be interpreted in the following way. Each line is numbered for reference purposes. Right arrows (-->) indicate departure of a TCP segment from TCP A to TCP B, or arrival of a segment at B from A. Left arrows (<--), indicate the reverse. Ellipsis (...) indicates a segment which is still in the network (delayed). An "XXX" indicates a segment which is lost or rejected. Comments appear in parentheses. TCP states represent the state AFTER the departure or arrival of the segment (whose contents are shown in the center of each line). Segment contents are shown in abbreviated form, with sequence number, control flags, and ACK field. Other fields such as window, addresses, lengths, and text have been left out in the interest of clarity.

Table: Basic 3-Way Handshake for Connection Synchronization.

| | TCP A | | TCP B |
|---|---|---|---|
| 1 | CLOSED | | LISTEN |
| 2 | SYN-SENT | --> <SEQ=100><CTL=SYN> | --> SYN-RECEIVED |
| 3 | ESTABLISHED | <-- <SEQ=300><ACK=101><CTL=SYN,ACK> | <-- SYN-RECEIVED |
| 4 | ESTABLISHED | --><SEQ=101><ACK=301><CTL=ACK> | --> ESTABLISHED |
| 5 | ESTABLISHED | --> <SEQ=101><ACK=301><CTL=ACK><DATA> | --> ESTABLISHED |

In line 2 of table, TCP A begins by sending a SYN segment indicating that it will use sequence numbers starting with sequence number 100. In line 3, TCP B sends a SYN and acknowledges the SYN it received from TCP A. Note that the acknowledgment field indicates TCP B is now expecting to hear sequence 101, acknowledging the SYN which occupied sequence 100.

At line 4, TCP A responds with an empty segment containing an ACK for TCP B's SYN; and in line 5, TCP A sends some data. Note that the sequence number of the segment in line 5 is the same as in line 4 because the ACK does not occupy sequence number space (if it did, we would wind up ACKing ACK's!).

Simultaneous initiation is only slightly more complex, as is shown in table. Each TCP cycles from CLOSED to SYN-SENT to SYN-RECEIVED to ESTABLISHED.

Table: Simultaneous Connection Synchronization.

| | TCP A | | TCP B |
|---|---|---|---|
| 1 | CLOSED | | CLOSED |
| 2 | SYN-SENT | --> <SEQ=100><CTL=SYN> | ... |
| 3 | SYN-RECEIVED | <--<SEQ=300><CTL=SYN> | <-- SYN-SENT |
| 4 | | ... <SEQ=100><CTL=SYN> | --> SYN-RECEIVED |
| 5 | SYN-RECEIVED | --> <SEQ=100><ACK=301><CTL=SYN,ACK> | ... |
| 6 | ESTABLISHED | <-- <SEQ=300><ACK=101><CTL=SYN,ACK> | <-- SYN-RECEIVED |
| 7 | | ...<SEQ=101><ACK=301><CTL=ACK> | --> ESTABLISHED |

The principle reason for the three-way handshake is to prevent old duplicate connection initiations from causing confusion. To deal with this, a special control message, reset, has been devised. If the receiving TCP is in a non-synchronized state (i.e., SYN-SENT, SYN-RECEIVED), it returns to LISTEN on receiving an acceptable reset. If the TCP is in one of the synchronized states (ESTABLISHED, FIN-WAIT-1, FIN-WAIT-2, CLOSE-WAIT, CLOSING, LAST-ACK, TIME-WAIT), it aborts the connection and informs its user. We discuss this latter case under "half-open" connections below.

Table: Recovery from Old Duplicate SYN.

| | TCP A | | TCP B |
|---|---|---|---|
| 1 | CLOSED | | LISTEN |
| 2 | SYN-SENT | --> <SEQ=100><CTL=SYN> | ... |
| 3 | (duplicate) | ...<SEQ=90><CTL=SYN> | --> SYN-RECEIVED |
| 4 | SYN-SENT | <-- <SEQ=300><ACK=91><CTL=SYN,ACK> | <-- SYN-RECEIVED |
| 5 | SYN-SENT | --> <SEQ=91><CTL=RST> | --> LISTEN |
| 6 | | ...<SEQ=100><CTL=SYN> | --> SYN-RECEIVED |
| 7 | SYN-SENT | <--<SEQ=400><ACK=301><CTL=ACK> | <-- SYN-RECEIVED |
| 8 | ESTABLISHED | --> <SEQ=101><ACK=401><CTL=ACK> | --> ESTABLISHED |

As a simple example of recovery from old duplicates, consider table. At line 3, an old duplicate SYN arrives at TCP B. TCP B cannot tell that this is an old duplicate, so it responds normally (line 4). TCP A detects that the ACK field is incorrect and returns a RST (reset) with its SEQ field selected to make the segment believable. TCP B, on receiving the RST, returns to the LISTEN state. When the original SYN (pun intended) finally arrives at line 6, the synchronization proceeds normally. If the SYN at line 6 had arrived before the RST, a more complex exchange might have occurred with RST's sent in both directions.

## Half-Open Connections and other Anomalies

An established connection is said to be "half-open" if one of the TCPs has closed or aborted the connection at its end without the knowledge of the other, or if the two ends of the connection have become desynchronized owing to a crash that resulted in loss of memory. Such connections will automatically become reset if an attempt is made to send data in either direction. However, half-open connections are expected to be unusual, and the recovery procedure is mildly involved.

If at site A the connection no longer exists, then an attempt by the user at site B to send any data on it will result in the site B TCP receiving a reset control message. Such a message indicates to the site B TCP that something is wrong, and it is expected to abort the connection.

Assume that two user processes A and B are communicating with one another when a crash occurs causing loss of memory to A's TCP. Depending on the operating system supporting A's TCP, it is likely that some error recovery mechanism exists. When the TCP is up again, A is likely to start again from the beginning or from a recovery point. As a result, A will probably try to OPEN the connection again or try to SEND on the connection it believes open. In the latter case, it receives the error message "connection not open" from the local (A's) TCP. In an attempt to establish the connection, A's TCP will send a segment containing SYN. This scenario leads to the example shown in table. After TCP A crashes, the user attempts to re-open the connection. TCP B, in the meantime, thinks the connection is open.

Table: Half-Open Connection Discovery.

| | TCP A | | TCP B |
|---|---|---|---|
| 1 | (CRASH) | | (send 300,receive 100) |
| 2 | CLOSED | | ESTABLISHED |
| 3 | SYN-SENT | --> <SEQ=400><CTL=SYN> | --> (??) |
| 4 | (!!) | <-- <SEQ=300><ACK=100><CTL=ACK> | <-- ESTABLISHED |
| 5 | SYN-SENT | --> <SEQ=100><CTL=RST> | --> (Abort!!) |
| 6 | SYN-SENT | | CLOSED |
| 7 | SYN-SENT | <-- <SEQ=400><CTL=ACK> | --> |

When the SYN arrives at line 3, TCP B, being in a synchronized state, and the incoming segment outside the window, responds with an acknowledgment indicating what sequence it next expects to hear (ACK 100). TCP A sees that this segment does not acknowledge anything it sent and, being unsynchronized, sends a reset (RST) because it has detected a half-open connection. TCP B aborts at line 5. TCP A will continue to try to establish the connection; the problem is now reduced to the basic 3-way handshake of table.

An interesting alternative case occurs when TCP A crashes and TCP B tries to send data on what it thinks is a synchronized connection. This is illustrated in table. In this case, the data arriving at TCP A from TCP B (line 2) is unacceptable because no such connection exists, so TCP A sends a RST. The RST is acceptable so TCP B processes it and aborts the connection.

Table: Active Side Causes Half-Open Connection Discovery.

| | TCP A | | TCP B |
|---|---|---|---|
| 1 | (CRASH) | | (send 300,receive 100) |
| 2 | (??) | <-- <SEQ=300><ACK=100><DATA=100><DATA=10><CTL=ACK | <-- ESTABLISHED |
| 3 | | --> <SEQ=100><CTL=RST> | --> (ABORT!!) |

Table: Old Duplicate SYN Initiates a Reset on two Passive Sockets.

| | TCP A | | TCP B |
|---|---|---|---|
| 1 | LISTEN | | LISTEN |
| 2 | | ... <SEQ=Z><CTL=SYN | --> SYN-RECEIVED |
| 3 | (??) | <-- <SEQ=X><ACK=Z+1><CTL=SYN,ACK | <-- SYN-RECEIVED |
| 4 | | --> <SEQ=Z+1><CTL=RST | --> (return to LISTEN!) |
| 5 | LISTEN | | LISTEN |

A variety of other cases are possible, all of which are accounted for by the following rules for RST generation and processing.

## Reset Generation

As a general rule, reset (RST) must be sent whenever a segment arrives which apparently is not intended for the current connection. A reset must not be sent if it is not clear that this is the case.

There are three groups of states:

- If the connection does not exist (CLOSED) then a reset is sent in response to any incoming segment except another reset. In particular, SYNs addressed to a non-existent connection are rejected by this means.
- If the incoming segment has an ACK field, the reset takes its sequence number from the ACK field of the segment, otherwise the reset has sequence number zero and the ACK field is set to the sum of the sequence number and segment length of the incoming segment. The connection remains in the CLOSED state.
- If the connection is in any non-synchronized state (LISTEN, SYN-SENT, SYN-RECEIVED), and the incoming segment acknowledges something not yet sent (the segment carries an unacceptable ACK), or if an incoming segment has a security level or compartment which does not exactly match the level and compartment requested for the connection, a reset is sent.

If our SYN has not been acknowledged and the precedence level of the incoming segment is higher than the precedence level requested then either raise the local precedence level (if allowed by the user and the system) or send a reset; or if the precedence level of the incoming segment is lower than the precedence level requested then continue as if the precedence matched exactly (if the remote TCP cannot raise the precedence level to match ours this will be detected in the next segment it sends, and the connection will be terminated then).

If our SYN has been acknowledged (perhaps in this incoming segment) the precedence level of the incoming segment must match the local precedence level exactly, if it does not a reset must be sent. If the incoming segment has an ACK field, the reset takes its sequence number from the ACK field of the segment, otherwise the reset has sequence number zero and the ACK field is set to the sum of the sequence number and segment length of the incoming segment. The connection remains in the same state.

If the connection is in a synchronized state (ESTABLISHED, FIN-WAIT-1, FIN-WAIT-2, CLOSE-WAIT, CLOSING, LAST-ACK, TIME-WAIT), any unacceptable segment (out of window sequence number or unacceptible acknowledgment number) must elicit only an empty acknowledgment segment containing the current send-sequence number and an acknowledgment indicating the next sequence number expected to be received, and the connection remains in the same state.

If an incoming segment has a security level, or compartment, or precedence which does not exactly match the level, and compartment, and precedence requested for the connection,a reset is sent and connection goes to the CLOSED state. The reset takes its sequence number from the ACK field of the incoming segment.

### Reset Processing

In all states except SYN-SENT, all reset (RST) segments are validated by checking their SEQ-fields. A reset is valid if its sequence number is in the window. In the SYN-SENT state (a RST received in response to an initial SYN), the RST is acceptable if the ACK field acknowledges the SYN.

The receiver of a RST first validates it, then changes state. If the receiver was in the LISTEN state, it ignores it. If the receiver was in SYN-RECEIVED state and had previously been in the LISTEN state, then the receiver returns to the LISTEN state, otherwise the receiver aborts the connection and goes to the CLOSED state. If the receiver was in any other state, it aborts the connection and advises the user and goes to the CLOSED state.

### Closing a Connection

CLOSE is an operation meaning "I have no more data to send." The notion of closing a full-duplex connection is subject to ambiguous interpretation, of course, since it may not be obvious how to treat the receiving side of the connection. We have chosen to treat CLOSE in a simplex fashion. The user who CLOSEs may continue to RECEIVE until he is told that the other side has CLOSED also. Thus, a program could initiate several SENDs followed by a CLOSE, and then continue to RECEIVE until signaled that a RECEIVE failed because the other side has CLOSED. We assume that the TCP will signal a user, even if no RECEIVEs are outstanding, that the other side has closed, so the user can terminate his side gracefully. A TCP will reliably deliver all buffers SENT before the connection was CLOSED so a user who expects no data in return need only wait to hear the connection was CLOSED successfully to know that all his data was received at the destination TCP. Users must keep reading connections they close for sending until the TCP says no more data.

There are essentially three cases:

- The user initiates by telling the TCP to CLOSE the connection.
- The remote TCP initiates by sending a FIN control signal.
- Both users CLOSE simultaneously.

### Case 1: Local user Initiates the Close

In this case, a FIN segment can be constructed and placed on the outgoing segment queue. No further SENDs from the user will be accepted by the TCP, and it enters the FIN-WAIT-1 state.

RECEIVEs are allowed in this state. All segments preceding and including FIN will be retransmitted until acknowledged. When the other TCP has both acknowledged the FIN and sent a FIN of its own, the first TCP can ACK this FIN. Note that a TCP receiving a FIN will ACK but not send its own FIN until its user has CLOSED the connection also.

## Case 2: TCP Receives a FIN from the Network

If an unsolicited FIN arrives from the network, the receiving TCP can ACK it and tell the user that the connection is closing. The user will respond with a CLOSE, upon which the TCP can send a FIN to the other TCP after sending any remaining data. The TCP then waits until its own FIN is acknowledged whereupon it deletes the connection. If an ACK is not forthcoming, after the user timeout the connection is aborted and the user is told.

## Case 3: Both users Close Simultaneously

A simultaneous CLOSE by users at both ends of a connection causes FIN segments to be exchanged. When all segments preceding the FINs have been processed and acknowledged, each TCP can ACK the FIN it has received. Both will, upon receiving these ACKs, delete the connection.

Table: Normal close sequence.

| | TCP A | | TCP B |
|---|---|---|---|
| 1 | ESTABLISHED | | ESTABLISHED |
| 2 | (Close) FIN-WAIT-1 | --> <SEQ=100><ACK=300><CTL=FIN, ACK> | --> CLOSE-WAIT |
| 3 | FIN-WAIT-2 | <-- <SEQ=300><ACK=101><CTL=ACK> | <-- CLOSE-WAIT |
| 4 | TIME-WAIT | <-- <SEQ=300><ACK=101><CTL=FIN,ACK> | (Close) |
| 5 | TIME-WAIT | <SEQ=101><ACK=301><CTL=ACK> | <-- LAST-ACK |
| 6 | (2 MSL) CLOSED | | --> CLOSED |

Table: Simultaneous close sequence.

| | TCP A | | TCP B |
|---|---|---|---|
| 1 | ESTABLISHED | | ESTABLISHED |
| 2 | (Close) FIN-WAIT-1 | --> <SEQ=100><ACK=300><CTL=FIN, ACK><br><--<SEQ=300><ACK=100><CTL=FIN, ACK><br>... <SEQ=100><ACK=300><CTL=FIN, ACK> | ...(Close) FIN-WAIT-1<br><--<br>--> |
| 3 | CLOSING | <-- <SEQ=101><ACK=301><CTL=ACK><br><-- <SEQ=301><ACK=101><CTL=ACK><br><-- <SEQ=101><ACK=301><CTL=ACK> | ... CLOSING<br><--<br>--> |
| 4 | TIME-WAIT<br>(2 MSL)<br>CLOSED | <-- <SEQ=300><ACK=101><CTL=FIN,ACK> | TIME-WAIT<br>(2 MSL)<br>CLOSED |

## Precedence and Security

The intent is that connection be allowed only between ports operating with exactly the same security and compartment values and at the higher of the precedence level requested by the two ports.

The precedence and security parameters used in TCP are exactly those defined in the Internet Protocol (IP). Throughout this TCP specification the term "security/compartment" is intended to indicate the security parameters used in IP including security, compartment, user group, and handling restriction.

A connection attempt with mismatched security/compartment values or a lower precedence value must be rejected by sending a reset. Rejecting a connection due to too low a precedence only occurs after an acknowledgment of the SYN has been received. TCP modules which operate only at the default value of precedence will still have to check the precedence of incoming segments and possibly raise the precedence level they use on the connection.

The security paramaters may be used even in a non-secure environment (the values would indicate unclassified data), thus hosts in non-secure environments must be prepared to receive the security parameters, though they need not send them.

### Data Communication

Once the connection is established data is communicated by the exchange of segments. Because segments may be lost due to errors (checksum test failure), or network congestion, TCP uses retransmission (after a timeout) to ensure delivery of every segment. Duplicate segments may arrive due to network or TCP retransmission. As discussed in the section on sequence numbers the TCP performs certain tests on the sequence and acknowledgment numbers in the segments to verify their acceptability.

The sender of data keeps track of the next sequence number to use in the variable SND.NXT. The receiver of data keeps track of the next sequence number to expect in the variable RCV.NXT. The sender of data keeps track of the oldest unacknowledged sequence number in the variable SND. UNA. If the data flow is momentarily idle and all data sent has been acknowledged then the three variables will be equal.

When the sender creates a segment and transmits it the sender advances SND.NXT. When the receiver accepts a segment it advances RCV.NXT and sends an acknowledgment. When the data sender receives an acknowledgment it advances SND.UNA. The extent to which the values of these variables differ is a measure of the delay in the communication. The amount by which the variables are advanced is the length of the data in the segment. Note that once in the ESTABLISHED state all segments must carry current acknowledgment information. The CLOSE user call implies a push function, as does the FIN control flag in an incoming segment.

### Retransmission Timeout

Because of the variability of the networks that compose an internetwork system and the wide range of uses of TCP connections the retransmission timeout must be dynamically determined. One procedure for determining a retransmission time out is given here as an illustration.

### An Example Retransmission Timeout Procedure

Measure the elapsed time between sending a data octet with a particular sequence number and receiving an acknowledgment that covers that sequence number (segments sent do not have to match segments received). This measured elapsed time is the Round Trip Time (RTT). Next compute a Smoothed Round Trip Time (SRTT) as:

SRTT = ( ALPHA * SRTT ) + ((1-ALPHA) * RTT)

and based on this, compute the retransmission timeout (RTO) as:

RTO = min[UBOUND,max[LBOUND,(BETA*SRTT)]]

where UBOUND is an upper bound on the timeout (e.g., 1 minute), LBOUND is a lower bound on the timeout (e.g., 1 second), ALPHA is a smoothing factor (e.g.,.8 to.9), and BETA is a delay variance factor.

## The Communication of Urgent Information

The objective of the TCP urgent mechanism is to allow the sending user to stimulate the receiving user to accept some urgent data and to permit the receiving TCP to indicate to the receiving user when all the currently known urgent data has been received by the user.

This mechanism permits a point in the data stream to be designated as the end of urgent information. Whenever this point is in advance of the receive sequence number (RCV.NXT) at the receiving TCP, that TCP must tell the user to go into "urgent mode"; when the receive sequence number catches up to the urgent pointer, the TCP must tell user to go into "normal mode". If the urgent pointer is updated while the user is in "urgent mode", the update will be invisible to the user.

The method employs a urgent field which is carried in all segments transmitted. The URG control flag indicates that the urgent field is meaningful and must be added to the segment sequence number to yield the urgent pointer. The absence of this flag indicates that there is no urgent data outstanding.

To send an urgent indication the user must also send at least one data octet. If the sending user also indicates a push, timely delivery of the urgent information to the destination process is enhanced.

## Managing the Window

The window sent in each segment indicates the range of sequence numbers the sender of the window (the data receiver) is currently prepared to accept. There is an assumption that this is related to the currently available data buffer space available for this connection.

Indicating a large window encourages transmissions. If more data arrives than can be accepted, it will be discarded. This will result in excessive retransmissions, adding unnecessarily to the load on the network and the TCPs. Indicating a small window may restrict the transmission of data to the point of introducing a round trip delay between each new segment transmitted.

The mechanisms provided allow a TCP to advertise a large window and to subsequently advertise a much smaller window without having accepted that much data. This, so called "shrinking the window," is strongly discouraged. The robustness principle dictates that TCPs will not shrink the window themselves, but will be prepared for such behavior on the part of other TCPs.

The sending TCP must be prepared to accept from the user and send at least one octet of new data even if the send window is zero. The sending TCP must regularly retransmit to the receiving TCP even when the window is zero. Two minutes is recommended for the retransmission interval when the window is zero. This retransmission is essential to guarantee that when either TCP has a zero window the re-opening of the window will be reliably reported to the other.

When the receiving TCP has a zero window and a segment arrives it must still send an acknowledgment showing its next expected sequence number and current window (zero).

The sending TCP packages the data to be transmitted into segments which fit the current window, and may repackage segments on the retransmission queue. Such repackaging is not required, but may be helpful.

In a connection with a one-way data flow, the window information will be carried in acknowledgment segments that all have the same sequence number so there will be no way to reorder them if they arrive out of order. This is not a serious problem, but it will allow the window information to be on occasion temporarily based on old reports from the data receiver. A refinement to avoid this problem is to act on the window information from segments that carry the highest acknowledgment number (that is segments with acknowledgment number equal or greater than the highest previously received).

The window management procedure has significant influence on the communication performance. The following comments are suggestions to implementers.

### Window Management Suggestions

Allocating a very small window causes data to be transmitted in many small segments when better performance is achieved using fewer large segments.

One suggestion for avoiding small windows is for the receiver to defer updating a window until the additional allocation is at least X percent of the maximum allocation possible for the connection (where X might be 20 to 40).

Another suggestion is for the sender to avoid sending small segments by waiting until the window is large enough before sending data. If the the user signals a push function then the data must be sent even if it is a small segment.

Note that the acknowledgments should not be delayed or unnecessary retransmissions will result. One strategy would be to send an acknowledgment when a small segment arrives (with out updating the window information), and then to send another acknowledgment with new window information when the window is larger.

The segment sent to probe a zero window may also begin a break up of transmitted data into smaller and smaller segments. If a segment containing a single data octet sent to probe a zero window is accepted, it consumes one octet of the window now available. If the sending TCP simply sends as much as it can whenever the window is non zero, the transmitted data will be broken into alternating big and small segments. As time goes on, occasional pauses in the receiver making window allocation available will result in breaking the big segments into a small and not quite so big pair. And after a while the data transmission will be in mostly small segments.

The suggestion here is that the TCP implementations need to actively attempt to combine small window allocations into larger windows, since the mechanisms for managing the window tend to lead to many small windows in the simplest minded implementations.

## Interfaces

There are of course two interfaces of concern: the user/TCP interface and the TCP/lower-level interface. We have a fairly elaborate model of the user/TCP interface, but the interface to the lower level protocol module is left unspecified here, since it will be specified in detail by the specification of the lowel level protocol. For the case that the lower level is IP we note some of the parameter values that TCPs might use.

## User/TCP Interface

The following functional description of user commands to the TCP is, at best, fictional, since every operating system will have different facilities. Consequently, we must warn readers that different TCP implementations may have different user interfaces. However, all TCPs must provide a certain minimum set of services to guarantee that all TCP implementations can support the same protocol hierarchy. This section specifies the functional interfaces required of all TCP implementations.

## TCP User Commands

Let us functionally characterize a USER/TCP interface. The notation used is similar to most procedure or function calls in high level languages, but this usage is not meant to rule out trap type service calls (e.g., SVCs, UUOs, EMTs).

The user commands described below specify the basic functions the TCP must perform to support interprocess communication. Individual implementations must define their own exact format, and may provide combinations or subsets of the basic functions in single calls. In particular, some implementations may wish to automatically OPEN a connection on the first SEND or RECEIVE issued by the user for a given connection.

In providing interprocess communication facilities, the TCP must not only accept commands, but must also return information to the processes it serves. The latter consists of:

- General information about a connection (e.g., interrupts, remote close, binding of unspecified foreign socket).
- Replies to specific user commands indicating success or various types of failure.

## Open

Format: OPEN (local port, foreign socket, active/passive [, timeout] [, precedence] [, security/compartment] [, options]) -> local connection name.

We assume that the local TCP is aware of the identity of the processes it serves and will check the authority of the process to use the connection specified. Depending upon the implementation of the TCP, the local network and TCP identifiers for the source address will either be supplied by the TCP or the lower level protocol (e.g., IP). These considerations are the result of concern about security, to the extent that no TCP be able to masquerade as another one, and so on. Similarly, no process can masquerade as another without the collusion of the TCP.

If the active/passive flag is set to passive, then this is a call to LISTEN for an incoming connection. A passive open may have either a fully specified foreign socket to wait for a particular connection or an unspecified foreign socket to wait for any call. A fully specified passive call can be made active by the subsequent execution of a SEND.

A transmission control block (TCB) is created and partially filled in with data from the OPEN command parameters.

On an active OPEN command, the TCP will begin the procedure to synchronize (i.e., establish) the connection at once.

The timeout, if present, permits the caller to set up a timeout for all data submitted to TCP. If data is not successfully delivered to the destination within the timeout period, the TCP will abort the connection. The present global default is five minutes.

The TCP or some component of the operating system will verify the users authority to open a connection with the specified precedence or security/compartment. The absence of precedence or security/compartment specification in the OPEN call indicates the default values must be used.

TCP will accept incoming requests as matching only if the security/compartment information is exactly the same and only if the precedence is equal to or higher than the precedence requested in the OPEN call.

The precedence for the connection is the higher of the values requested in the OPEN call and received from the incoming request, and fixed at that value for the life of the connection. Implementers may want to give the user control of this precedence negotiation. For example, the user might be allowed to specify that the precedence must be exactly matched, or that any attempt to raise the precedence be confirmed by the user.

A local connection name will be returned to the user by the TCP. The local connection name can then be used as a short hand term for the connection defined by the pair.

## Send

Format: SEND (local connection name, buffer address, byte count, PUSH flag, URGENT flag [,timeout]).

This call causes the data contained in the indicated user buffer to be sent on the indicated connection. If the connection has not been opened, the SEND is considered an error. Some implementations may allow users to SEND first; in which case, an automatic OPEN would be done. If the calling process is not authorized to use this connection, an error is returned.

If the PUSH flag is set, the data must be transmitted promptly to the receiver, and the PUSH bit will be set in the last TCP segment created from the buffer. If the PUSH flag is not set, the data may be combined with data from subsequent SENDs for transmission efficiency.

If the URGENT flag is set, segments sent to the destination TCP will have the urgent pointer set. The receiving TCP will signal the urgent condition to the receiving process if the urgent pointer indicates that data preceding the urgent pointer has not been consumed by the receiving process. The purpose of urgent is to stimulate the receiver to process the urgent data and to indicate to the receiver when all the currently known urgent data has been received. The number of times the sending user's TCP signals urgent will not necessarily be equal to the number of times the receiving user will be notified of the presence of urgent data.

If no foreign socket was specified in the OPEN, but the connection is established (e.g., because a LISTENing connection has become specific due to a foreign segment arriving for the local socket), then the designated buffer is sent to the implied foreign socket. Users who make use of OPEN with an unspecified foreign socket can make use of SEND without ever explicitly knowing the foreign socket address.

However, if a SEND is attempted before the foreign socket becomes specified, an error will be returned. Users can use the STATUS call to determine the status of the connection. In some implementations the TCP may notify the user when an unspecified socket is bound. If a timeout is specified, the current user timeout for this connection is changed to the new one.

In the simplest implementation, SEND would not return control to the sending process until either the transmission was complete or the timeout had been exceeded. However, this simple method is both subject to deadlocks (for example, both sides of the connection might try to do SENDs before doing any RECEIVEs) and offers poor performance, so it is not recommended. A more sophisticated implementation would return immediately to allow the process to run concurrently with network I/O, and, furthermore, to allow multiple SENDs to be in progress. Multiple SENDs are served in first come, first served order, so the TCP will queue those it cannot service immediately.

We have implicitly assumed an asynchronous user interface in which a SEND later elicits some kind of SIGNAL or pseudo-interrupt from the serving TCP. An alternative is to return a response immediately. For instance, SENDs might return immediate local acknowledgment, even if the segment sent had not been acknowledged by the distant TCP. We could optimistically assume eventual success. If we are wrong, the connection will close anyway due to the timeout. In implementations of this kind (synchronous), there will still be some asynchronous signals, but these will deal with the connection itself, and not with specific segments or buffers.

In order for the process to distinguish among error or success indications for different SENDs, it might be appropriate for the buffer address to be returned along with the coded response to the SEND request. TCP-to-user signals are discussed below, indicating the information which should be returned to the calling process.

### Receive

Format: RECEIVE (local connection name, buffer address, byte count) -> byte count, urgent flag, push flag.

This command allocates a receiving buffer associated with the specified connection. If no OPEN precedes this command or the calling process is not authorized to use this connection, an error is returned.

In the simplest implementation, control would not return to the calling program until either the buffer was filled, or some error occurred, but this scheme is highly subject to deadlocks. A more sophisticated implementation would permit several RECEIVEs to be outstanding at once. These would be filled as segments arrive. This strategy permits increased throughput at the cost of a more elaborate scheme (possibly asynchronous) to notify the calling program that a PUSH has been seen or a buffer filled.

If enough data arrive to fill the buffer before a PUSH is seen, the PUSH flag will not be set in the response to the RECEIVE. The buffer will be filled with as much data as it can hold. If a PUSH is seen before the buffer is filled the buffer will be returned partially filled and PUSH indicated. If there is urgent data the user will have been informed as soon as it arrived via a TCP-to-user signal. The receiving user should thus be in "urgent mode". If the URGENT flag is on, additional urgent data remains. If the URGENT flag is off, this call to RECEIVE has returned all the urgent data, and the user may now leave "urgent mode". Note that data following the urgent pointer (non-urgent data) cannot be delivered to the user in the same buffer with preceeding urgent data unless the boundary is clearly marked for the user.

To distinguish among several outstanding RECEIVEs and to take care of the case that a buffer is not completely filled, the return code is accompanied by both a buffer pointer and a byte count indicating the actual length of the data received.

Alternative implementations of RECEIVE might have the TCP allocate buffer storage, or the TCP might share a ring buffer with the user.

## Close

Format: CLOSE (local connection name): This command causes the connection specified to be closed. If the connection is not open or the calling process is not authorized to use this connection, an error is returned. Closing connections is intended to be a graceful operation in the sense that outstanding SENDs will be transmitted (and retransmitted), as flow control permits, until all have been serviced. Thus, it should be acceptable to make several SEND calls, followed by a CLOSE, and expect all the data to be sent to the destination. It should also be clear that users should continue to RECEIVE on CLOSING connections, since the other side may be trying to transmit the last of its data. Thus, CLOSE means "I have no more to send" but does not mean "I will not receive any more." It may happen (if the user level protocol is not well thought out) that the closing side is unable to get rid of all its data before timing out. In this event, CLOSE turns into ABORT, and the closing TCP gives up.

The user may CLOSE the connection at any time on his own initiative, or in response to various prompts from the TCP (e.g., remote close executed, transmission timeout exceeded, destination inaccessible).

Because closing a connection requires communication with the foreign TCP, connections may remain in the closing state for a short time. Attempts to reopen the connection before the TCP replies to the CLOSE command will result in error responses. Close also implies push function.

## Status

Format: STATUS (local connection name) -> status data

This is an implementation dependent user command and could be excluded without adverse effect. Information returned would typically come from the TCB associated with the connection. This command returns a data block containing the following information:

- local socket,
- foreign socket,
- local connection name,
- receive window,
- send window,
- connection state,
- number of buffers awaiting acknowledgment,
- number of buffers pending receipt,
- urgent state,
- precedence,
- security/compartment,
- transmission timeout.

Depending on the state of the connection, or on the implementation itself, some of this information may not be available or meaningful. If the calling process is not authorized to use this connection, an error is returned. This prevents unauthorized processes from gaining information about a connection.

## Abort

Format: ABORT (local connection name): This command causes all pending SENDs and RECEIVES to be aborted, the TCB to be removed, and a special RESET message to be sent to the TCP on the other side of the connection. Depending on the implementation, users may receive abort indications for each outstanding SEND or RECEIVE, or may simply receive an ABORT-acknowledgment.

## TCP-to-user Messages

It is assumed that the operating system environment provides a means for the TCP to asynchronously signal the user program. When the TCP does signal a user program, certain information is passed to the user. Often in the specification the information will be an error message. In other cases there will be information relating to the completion of processing a SEND or RECEIVE or other user call. The following information is provided:

| Local Connection Name | Always |
|---|---|
| Response String | Always |
| Buffer Address | Send & Receive |
| Byte count (counts bytes received) | Receive |

| Push flag | Receive |
|---|---|
| Urgent flag | Receive |

## TCP/Lower-Level Interface

The TCP calls on a lower level protocol module to actually send and receive information over a network. One case is that of the ARPA internetwork system where the lower level module is the Internet Protocol (IP).

If the lower level protocol is IP it provides arguments for a type of service and for a time to live. TCP uses the following settings for these parameters:

- Type of Service = Precedence: routine, Delay: normal, Throughput: normal, Reliability: normal; or 00000000.
- Time to Live = one minute, or 00111100.

Note that the assumed maximum segment lifetime is two minutes. Here we explicitly ask that a segment be destroyed if it cannot be delivered by the internet system within one minute.

If the lower level is IP (or other protocol that provides this feature) and source routing is used, the interface must allow the route information to be communicated. This is especially important so that the source and destination addresses used in the TCP checksum be the originating source and ultimate destination. It is also important to preserve the return route to answer connection requests.

Any lower level protocol will have to provide the source address, destination address, and protocol fields, and some way to determine the "TCP length", both to provide the functional equivlent service of IP and to be used in the TCP checksum.

## Event Processing

The processing depicted is an example of one possible implementation. Other implementations may have slightly different processing sequences, but they should differ from those in this section only in detail, not in substance.

The activity of the TCP can be characterized as responding to events. The events that occur can be cast into three categories: user calls, arriving segments, and timeouts. This section describes the processing the TCP does in response to each of the events. In many cases the processing required depends on the state of the connection. Events that occur:

### User Calls

- OPEN
- SEND
- RECEIVE
- CLOSE
- ABORT
- STATUS

## Arriving Segments

- SEGMENT ARRIVES

## Timeouts

- USER TIMEOUT
- RETRANSMISSION TIMEOUT
- TIME-WAIT TIMEOUT

The model of the TCP/user interface is that user commands receive an immediate return and possibly a delayed response via an event or pseudo interrupt. In the following descriptions, the term "signal" means cause a delayed response.

Error responses are given as character strings. For example, user commands referencing connections that do not exist receive "error: connection not open".

Please note in the following that all arithmetic on sequence numbers, acknowledgment numbers, windows, et cetera, is modulo 2**32 the size of the sequence number space. Also note that "=<" means less than or equal to (modulo 2**32).

A natural way to think about processing incoming segments is to imagine that they are first tested for proper sequence number (i.e., that their contents lie in the range of the expected "receive window" in the sequence number space) and then that they are generally queued and processed in sequence number order.

When a segment overlaps other already received segments we reconstruct the segment to contain just the new data, and adjust the header fields to be consistent. If no state change is mentioned the TCP stays in the same state.

5

# Cellular Networks

A cellular network is a mobile network that provides services over land through cells and base stations. Cellular networks provide a larger coverage area with increased capacity over alternative solutions. Cellular network technology has come a long way from 2G to 3G and 4G with 5G being the latest introduction. This chapter closely examines the various cellular networks to provide an extensive understanding of the subject.

Cellular networks are high-speed, high-capacity voice and data communication networks with enhanced multimedia and seamless roaming capabilities for supporting cellular devices. With the increase in popularity of cellular devices, these networks are used for more than just entertainment and phone calls. They have become the primary means of communication for finance-sensitive business transactions, lifesaving emergencies, and life-/ mission-critical services such as E-911. Today these networks have become the lifeline of communications.

A breakdown in a cellular network has many adverse effects, ranging from huge economic losses due to financial transaction disruptions; loss of life due to loss of phone calls made to emergency workers; and communication outages during emergencies such as the September 11, 2001, attacks. Therefore, it is a high priority for cellular networks to function accurately.

It must be noted that it is not difficult for unscrupulous elements to break into a cellular network and cause outages. The major reason for this is that cellular networks were not designed with security in mind. They evolved from the old-fashioned telephone networks that were built for performance. To this day, cellular networks have numerous well-known and unsecured vulnerabilities providing access to adversaries. Another feature of cellular networks is network relationships (also called dependencies) that cause certain types of errors to propagate to other network locations as a result of regular network activity. Such propagation can be very disruptive to a network, and in turn it can affect subscribers. Finally, Internet connectivity to cellular networks is another major contributor to cellular networks' vulnerability because it gives Internet users direct access to cellular network vulnerabilities from their homes.

To ensure that adversaries do not access cellular networks and cause breakdowns, a high level of security must be maintained in cellular networks. However, though great efforts have been made to improve cellular networks in terms of support for new and innovative services, greater number of subscribers, higher speed, and larger bandwidth, very little has been done to update the security of cellular networks. Accordingly, these networks have become highly attractive targets to adversaries, not only because of their lack of security but also due to the ease with which these networks can be exploited to affect millions of subscribers.

For a cellular network, the base stations (BS), along with the radio network controllers (RNC), are equipped with IT capabilities, and thus edge analytics would enable us to track real-time user context to enhance service delivery and improve overall QoE for the end users. In this context, we define.

## Edge

This refers to the basestations (BS), or the (distributed) wireless system comprising the BS, to which a user connects. The edge is restricted in its storage capacity and computational capability. The primary role of the edge lies in performing fast online or incremental analytics.

## Core

This refers to the central backbone components of the network comprising RNC, SGSN, and GGSN, and is considered to be endowed with practically infinite storage and computation resources for performing rigorous batch mode or offline analytics.

The edge takes part in the real-time handling of users (device hand-off, etc.) and other analytical decisions. The core, bestowed with greater storage and computing power, usually works in an offline fashion, receives online user tracking from the basestations, updates the historical database correspondingly, and executes a rigorous frequent-pattern-mining, association-rule-mining algorithm to generate the frequent rules and behaviors on a per-user basis. These rules are later queried by the edges for subsequent online processing (eg, ranking and prefetching). It also handles housing, log generation, and updating the databases and performing queries. Thus, an efficient system evolves as a result of interplay between the online and offline computations.

The combination of frequent pattern mining, prediction, and aggregation of the results forms the basis of the INCA algorithm and provides a novel approach to efficiently tackle the expanse of such network-based services. The proposed system can cater not only to the VoD domain, but also other wireless network-based applications such as advertisement insertion, location-based services, etc., thereby making it a general framework for a broad range of scenarios requiring network intelligence and analytics at the heart.

A typical VoD scenario involves a user request to the basestation. If the requested video is available, it is directly streamed. Otherwise, the basestation initiates a request to the core. The video is then fed to the basestation from where it is provided to the users. Figure depicts a simple VoD request scenario.

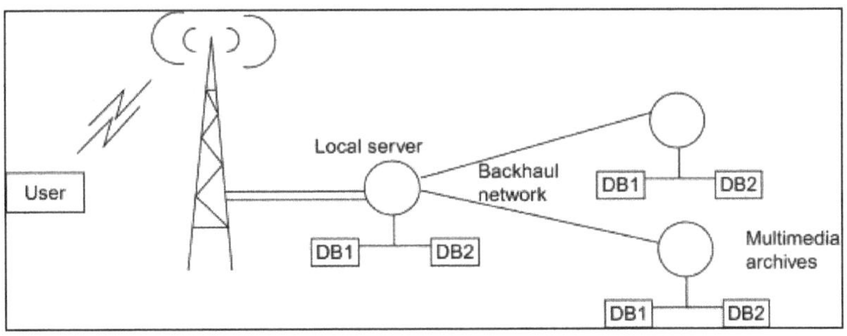


Interactive VoD services can be broadly classified into the following categories:

- Broadcast (no VoD).
- Pay-per-view.
- Quasi VoD.

- Near VoD.
- True Video-on-Demand (T-VoD).

## Zigbee Networks

ZigBee is a technological standard designed for control and sensor networks. It is based on the IEEE 802.15.4 standard, and has been created by the ZigBee Alliance to support reliable, cost-effective, low power wireless applications. It operates in Personal Area Networks (PAN's) and device-to-device networks. It provides connectivity between small packet devices, and is meant for control of lights, switches, thermostats, appliances, etc. It has applications in industrial and commercial equipment for monitoring and automated control of sensors and actuators, consumer electronics, toys and games, home automation, pc-peripheral control etc. The characteristics of ZigBee are:

- Low cost.
- Low power consumption.
- Low data rate.
- Relatively short transmission range.
- Scalability.
- Reliability.
- Flexible protocol design suitable for many applications.

### ZigBee/Ieee 802.15.4 Stack

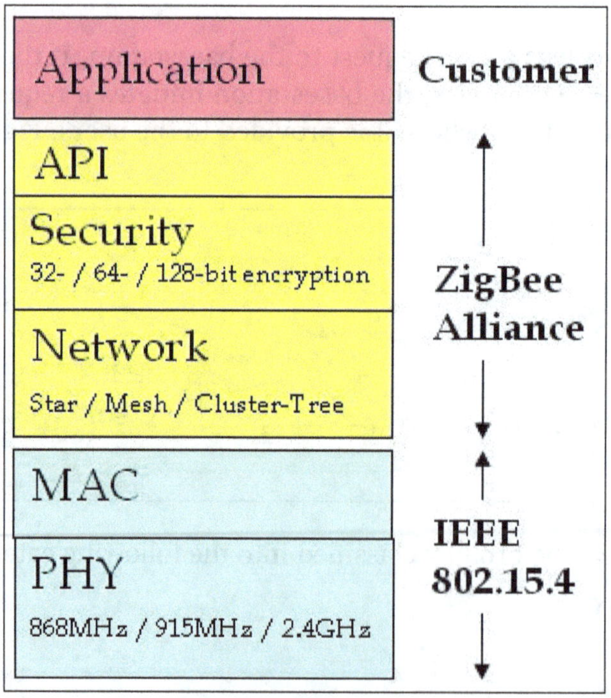


Figure. ZigBee Protocol Stack.

The protocol stack shown in Fig can be viewed as being controlled by three players. The Physical and MAC layers are defined by the IEEE 802.15.4 standard. The network, security and API standards, and application profiles are specified by the ZigBee Alliance, and the application is handled by the customer/user. The IEEE 802.15.4 component is handled by the hardware, whereas the network, security and application layers are handled by software. Support for security is one of the key features of ZigBee.

### IEEE 802.15.4 Physical Layer

The functionalities of the PHY layer include:

- Activation and deactivation of the radio transceiver.
- Energy detection within the current channel.
- Link quality indication for received packets.
- Clear channel assessment for CSMA-CA.
- Channel frequency selection.
- Data transmission and reception.

The frame structure of the physical layer is given below:

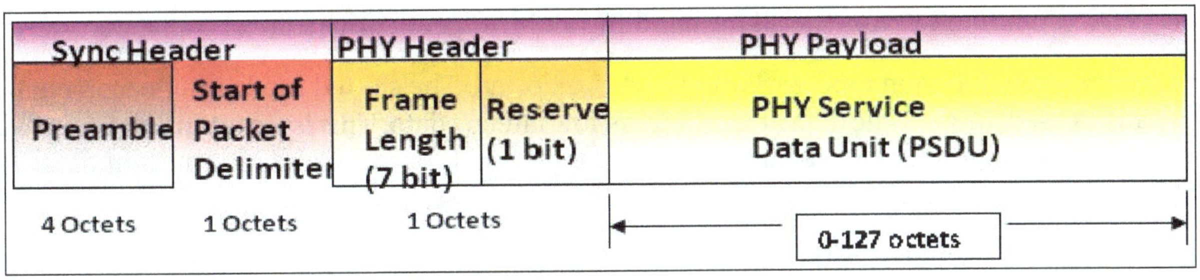


It starts with a Preamble field of (32 bits), which is used for synchronization. This is followed by a "Start of packet delimiter" (8 bits) which shall be formatted as "11100101". These two fields constitute the Sync Header. Next comes the PHY header which gives the length of the payload in 8 bits. Following this is the payload – PHY Service Data Unit PSDU which can be 0 to 127 bytes long.

The standard specifies two PHYs – two operating frequency bands are:

- 868 MHz/915 MHz direct sequence spread spectrum (DSSS) PHY ( with11 channels). Of these 1 channel (20Kb/s) is in the European 868MHz band 10 channels (40Kb/s) are in the 915 (902-928)MHz ISM band.

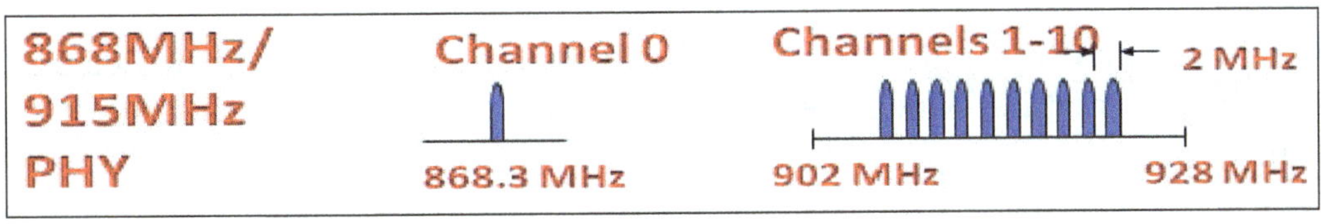


- 2450 MHz (2.4GHz) direct sequence spread spectrum (DSSS) PHY (with16 channels of 250Kb/s) in 2.4GHz band.

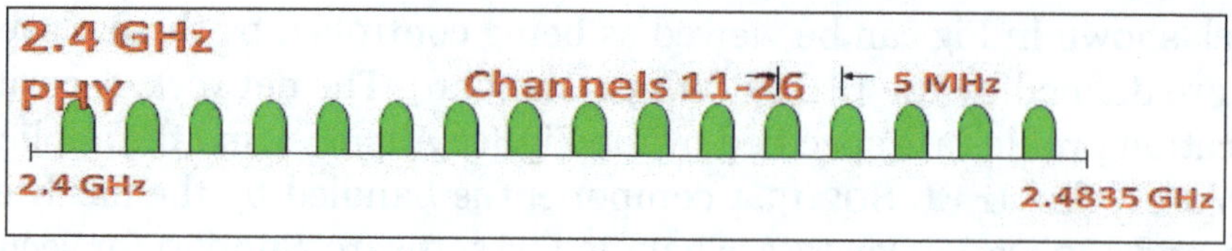


## IEEE 802.15.4 MAC Layer

CSMA/CA is the access method used in IEEE 802.15.4. An optional slotted mechanism is also supported. The MAC layer frame structure is designed to support 3 different types of traffic:

- Periodic data as in data from sensors.
- Intermittent data as in data from a light switch.
- Repetitive low latency data as from a mouse.

There are different frame types provided in 802.15.4. They are:

- Data Frame - used for all transfers of data.
- Beacon Frame - used by a coordinator to transmit beacons.
- Acknowledgment Frame - used for confirming successful frame reception.
- MAC Command Frame - used for handling all MAC peer entity control transfers.

There are two transmission modes: Slotted and Un-slotted. The slotted mode is a Beacon enabled mode, and is used for periodic data and repetitive low latency data. This is a contention free period. The Un-slotted mode is for Non-Beacon enabled mode, and is used to send intermittent data. This is a contention access period.

There are two classes of devices supported: Full function device (FFD), and Reduced function device (RFD).

FFDs can function in any topology, and can talk to any other device (FFD/RFD). It is capable of being a Network coordinator, and a PAN coordinator. RFDs on the other hand are limited to star topology, and can talk only to FFDs. They cannot become network coordinators. They can just be end devices.

All devices must have 64 bit IEEE addresses, but short (16 bit) addresses can be allocated by the coordinator to reduce the packet size.

## ZigBee Network Topologies

There are three network topologies supported by ZigBee – star, mesh and cluster tree. The star topology is a simple structure, which is easy to synchronize with low latencies. But it is limited to small networks.

The mesh topology supports robust multi-hop communication, and the network is more flexible. It has lower latency. But routing function is to be supported – the route discovery is costly, and we need storage for the routing table at the nodes. The Cluster tree is actually a cluster of trees. It has the advantage of low routing cost and support for multi-hop communication. But route reconstruction is costly, and the latency may be quite long.

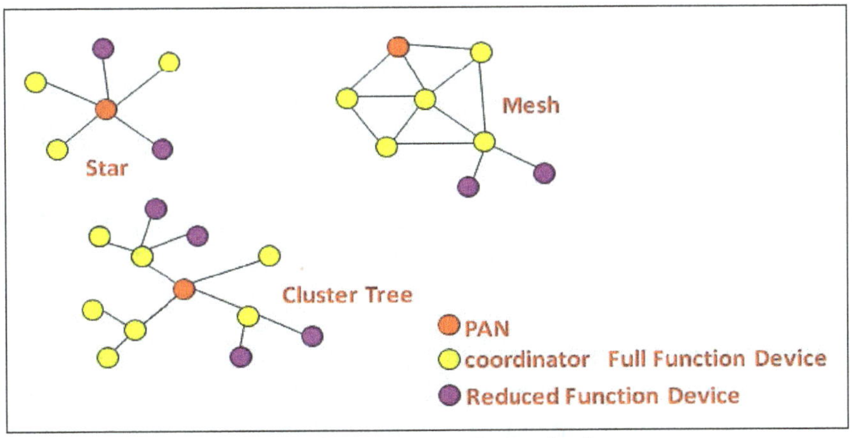


Figure: ZigBee Network Topologies.

## ZigBee vs Bluetooth

With this overview of ZigBee, it is interesting to compare it with the Bluetooth protocol. They are optimized for different applications. ZigBee is meant to carry smaller packets over a large network; while Bluetooth is meant for carrying larger packets over a small network. Zigee mostly caters to static networks with many, infrequently used devices, as in home automation, toys, remote controls, etc.; whereas Bluetooth is for Ad-hoc networks to support applications such as file transfer, screen graphics, pictures, hands-free audio, Mobile phones, headsets, PDAs, etc. A comparison of the two in terms of other features is given below in table.

Table: ZigBee and Bluetooth Comparison.

| Feature(s) | Bluetooth | ZigBee |
|---|---|---|
| Power Profile | Days | years |
| Complexity | Complex | Simple |
| Nodes/Master | 7 | 64000 |
| Latency | 10 seconds | 30 ms – 1s |
| Range | 10m | 70m ~ 300m |
| Extendibility | no | Yes |
| Data Rate | 1 Mbps | 250 Kbps |
| Security | 64bit, 128bit | 128bit AES and Application Layer |

# 2G Network

Technology is moving at an incredibly fast rate, and you don't have to be a tech-savvy person to reap the benefits of this. Many of the services you rely on everyday use this technology and while you don't have to understand it, it helps to be informed of how it affects you.

When technologies become outdated and replaced by better, faster technologies, it means that new services become available, or current services become better. One downside, though, is that if you are using outdated technology, you may experience poor service that can cause you problems, sometimes serious ones.

When you hear the term "2G" you may first think of your phone service carrier. You would be correct. AT&T and TMobile both use 2G, 3G and 4G networks to provide mobile service to their customers. This bandwidth was the first of its kind and replaced the older, analog technology several years ago. While analog is still in use, of course, a majority of people rely on 2, 3 and 4G services for their mobile phones, tablets, computers and more. The "G" simply stands for Generation.

In addition to phone service, though, there are other industries that use these networks. One of them may be your home security system. As far as mobile phones are concerned, 2G is fairly outdated, having been virtually replaced by faster and more advanced 3G and 4G networks that you likely use every day. 2G is still alive and well, however, and is relied on by many businesses.

All this is about to change in the next few years. The decision has been made to permanently retire 2G in order to free up network "space" needed by the faster networks. Over time, 2G will simply cease to exist.

### How does this affect your Security System?

Many security systems rely on AT&T's 2G network for their radios to operate, to communicate between the customers's home, the security headquarters and police and fire stations. Without the 2G network, service will cease to exist.

Of course, the shutdown is still a few years away, giving home security companies plenty of time to switch their networks over to the faster and more reliable 3G or 4G systems. However, even now towers are being converted to 3G and 4G to increase their service area. This means that resources are being pulled away from 2G in order to support the faster networks.

### How does this affect you Right Now?

The problem is, even before the shutdown, networks will begin to reroute and redistribute bandwidth, lowering the amount available for the older networks. What this means is that customers who use and rely on 2G will experience delays and drops in their service.

If your alarm company is currently running on this bandwidth, then there is a good chance their service is already being compromised. Their radios may already be experiencing slowed down response times or crowded networks. This means that your security service may fail you when you need it most.

### So what should you do?

Before a problem happens, you should consult with your current home security system to find out if they have made the switch to 3G or 4G service. If they have, then you don't have anything to worry about. If they haven't, find out if they are in the process of making this change. Again, the final shutdown will not be taking place for about three to five years, so some companies may be waiting until the deadline is a bit closer in order to avoid the costs of making the switch. The danger here is that their service may be affected.

If you are informed that your home security provider is in the process of switching their radios to a higher bandwidth, be sure to ask them how long this will take, and if there will be any additional costs to their service.

If you find that your home security service is using the out-dated bandwidth, it is best to switch over to a company that is using at least 3G services. By doing this, you are ensuring that your service does not experience any delays in the event of an emergency. For example, if someone breaks into your home, you want to know that the message gets to the authorities immediately, so that a car can be dispatched. A delay of even thirty seconds can make a huge difference should someone be breaking into your home.

## GSM

Digital cellular technology like GSM (Global System for Mobile Communication) is used to transmit mobile data as well as voice services. This concept was implemented at Bell Laboratories using a mobile radio system in 1970. As the name suggests, it is the standardization group name that was established in the year 1982 to make a general European mobile telephone standard. This technology owns above 70% of the market share of the digital cellular subscriber around the world. This technology was developed by using digital technology. At present, GSM technology supports above 1 billion mobile subscribers around the world in the above 210 countries. This technology provides voice and data services from fundamental to complex.

Figure: GSM Modem.

GSM is a mobile communication modem; it stands for global system for mobile communication (GSM). The idea of GSM was developed at Bell Laboratories in 1970. It is a widely used mobile communication system in the world. GSM is an open and digital cellular technology used for transmitting mobile voice and data services operate at the 850MHz, 900MHz, 1800MHz, and 1900MHz frequency bands.

GSM technology was developed as a digital system using the time division multiple access (TDMA) technique for communication purposes. A GSM digitizes and reduces the data, then sends it down through a channel with two different streams of client data, each in its own particular time slot. The digital system has the ability to carry 64 kbps to 120 Mbps of data rates.

There are various cell sizes in a GSM system such as macro, micro, pico, and umbrella cells. Each cell varies as per the implementation domain. There are five different cell sizes in a GSM network macro, micro, pico, and umbrella cells. The coverage area of each cell varies according to the implementation environment.

The time division multiple access (TDMA) technique relies on assigning different time slots to each user on the same frequency. It can easily adapt to data transmission and voice communication and can carry 64 kbps to 120 Mbps of data rate.

### Features of GSM Module

The features of the GSM module include the following:

- Improved spectrum efficiency.
- International roaming.
- Compatibility with integrated services digital network (ISDN).
- Support for new services.
- SIM phonebook management.
- Fixed dialing number (FDN).
- Real-time clock with alarm management.
- High-quality speech.
- Uses encryption to make phone calls more secure.
- Short message service (SMS).

The security strategies standardized for the GSM system make it the most secure telecommunications standard currently accessible. Although the confidentiality of a call and secrecy of the GSM subscriber is just ensured on the radio channel, this is a major step in achieving end-to-end security.

## GSM Modem

A GSM modem is a device that can be either a mobile phone or a modem device that can be used to make a computer or any other processor communicate over a network. A GSM modem requires a SIM card to be operated and operates over a network range subscribed by the network operator. It can be connected to a computer through serial, USB, or Bluetooth connection.

A GSM modem can also be a standard GSM mobile phone with the appropriate cable and software driver to connect to a serial port or USB port on your computer. GSM modem is usually preferable to a GSM mobile phone. The GSM modem has a wide range of applications in transaction terminals, supply chain management, security applications, weather stations, and GPRS mode remote data logging.

## Working of GSM Module

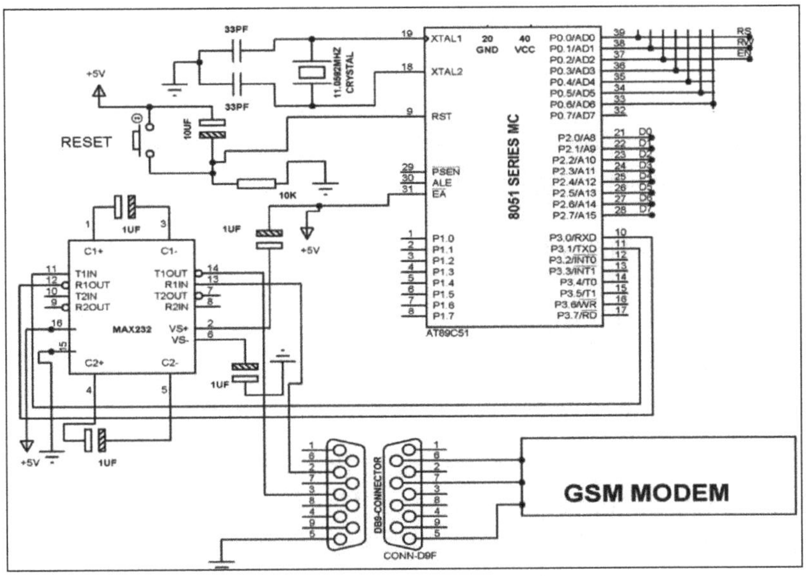


Figure: GMS Modem Circuit.

From the circuit, a GSM modem duly interfaced to the MC through the level shifter IC Max232. The SIM card mounted GSM modem upon receiving digit command by SMS from any cell phone sends that data to the MC through serial communication. While the program is executed, the GSM modem receives the command 'STOP' to develop an output at the MC, the contact point of which are used to disable the ignition switch.

The command so sent by the user is based on an intimation received by him through the GSM modem 'ALERT' a programmed message only if the input is driven low. The complete operation is displayed over a 16×2 LCD display.

## GSM Technology Applications

The applications of GSM technology include the following:

### Intelligent GSM Technology for Automation and Security

These days, the GSM mobile terminal has become one of the items that are constantly with us. Just like our wallet/purse, keys or watch, the GSM mobile terminal provides us a communication channel that enables us to communicate with the world. The requirement for a person to be reachable or to call anyone at any time is very appealing.

This project, as the name says the project is based on GSM network technology for transmission of SMS from sender to receiver. SMS sending and receiving is used for ubiquitous access to appliances and allowing breach control at home. The system proposes two sub-systems. The appliance control subsystem enables the user to control home appliances remotely and the security alert subsystem gives automatic security monitoring.

The system is capable enough to instruct users via SMS from a specific cell number to change the condition of the home appliance according to the user's needs and requirements. The second aspect is that of security alert which is achieved in a way that on the detection of intrusion, the system allows automatic generation of SMS thus alerting the user against security risk.

GSM technology will allow communication anywhere, anytime, and with anyone. The functional architecture of GSM employing intelligent networking principles, and its ideology, which provides the development of GSM is the first step towards a true personal communication system that enough standardization to ensure compatibility.

## GSM Applications in Medical Services

Consider two situations like the following:

- A person is critically injured or has fallen ill and needs to be immediately taken care of. All he or the person accompanying him has is a mobile phone.
- A patient is discharged from the hospital and thinks of taking rest at his home, but still has to go to the hospital for regular check-ups. He may have a mobile phone and also some medical sensor devices like health monitoring devices.

In both situations, the only way which can provide a solution is by using the mobile communication system. In other words, using communication technologies any situation like the above can be handled just by transmitting the patient details through the communication network and receiving them and processing them at the receiver section-either a the healthcare center or at the doctor's home.

The doctor simply monitors the patient details and gives back the instructions to the person(in the 1st case)so that he can at least take some precautions before finally reaching the hospital and in the 2nd case monitors the test results of the patient and in case of any abnormalities, takes the next step for further treatment. This whole situation is the telemedicine services. The telemedicine system can be used in any of the three ways.

- Using video conferencing, where patients sitting at one place can have direct interaction with the health care providers and accordingly carry on the curing process.
- By using health monitoring sensors that keep updating about the health of the patient and accordingly guide the health care providers to carry on the treatment.
- By transmitting the acquired medical data and transmit the acquired data for consultation and processing.

For the above three ways, a wireless communication technique is used. Medical services require many ways of getting access to stored resources. These can be medical databases or online hosts with devices that can help recover and monitor patient health. Different access options are broadband network, through medium-throughput media and narrowband through GSM. The advantages of GSM technology in the telemedicine system include the following:

- It is more cost-effective.
- GSM receivers are widely available- mobile phones and GSM modems.
- It has a high data transfer speed.

The architecture of a 2G GSM network is shown in Fig. Each cell has a Base Station or Base Transceiver Station (BTS). Multiple Cells come under the control of a Base Station controller (BSC or BS for short); and form a Base station system (BSS). Many BSCs are connected to a Mobile Switching Centre (MSC). Many MSCs connect to the public network through a Gateway MSC. The MTSC to BS link may be wired or wireless.

The BS which is at the centre of each cell, has an antenna, transceivers, and controller. The controller in the BS handles the call process. The MTSC is responsible for:

- Connecting calls between mobile units and from mobile to fixed telecommunications network.
- Assigning voice channels.
- Performing handoffs.
- Monitoring calls (billing).

When a mobile unit enters a cell area, it contacts the BS, which in turn passes the information to the MSC. The MSC then takes the responsibility of facilitating calls to and from this mobile unit.

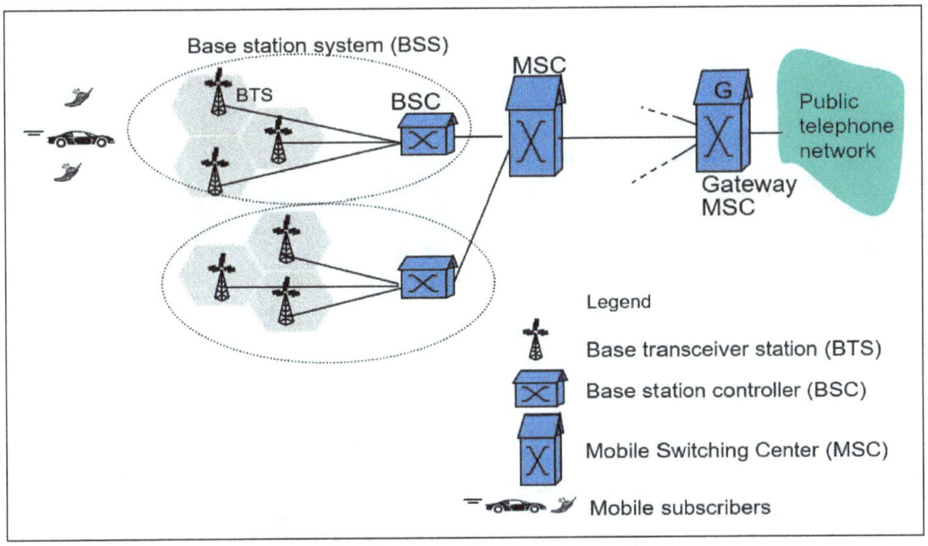


Figure: 2G GSM network architecture.

The channels using which the BS talks to the mobile units are divided in to two categories – "control" channels and "traffic" channels. The control channels are used for setting up and maintaining calls, and establishes the relationship between the mobile unit and the nearest BS. The traffic channels carry voice and data. The traffic channels are assigned only when a call is to be completed or data is to be transferred.

## Handling Mobility – Roaming User

In order to understand how mobility is handled, we need to know some terms that are used in this context. Home network is the network of the cellular provider that you subscribe to (e.g., Airtel, BSNL, etc.).

Home Location Register (HLR) is the database in the home network's MSC containing permanent cell phone #, profile information (services, preferences, billing), information about current location (could be in another network). Visited network is the network in which the mobile currently resides. Visitor Location Register (VLR) is the database with an entry for each user currently in the network. It could be the home network as well.

When a mobile enters a cell, its location (the BS it is attached to) is stored in the VLR of the corresponding MSC. If it is the home network, nothing else needs to be done. When a call comes to this device, it is routed to the home MSC from the public switched network. The home MSC looks at the VLR, identifies the BS to which the mobile is connected, and assigns a (traffic) channel using which this device can communicate the data with the BS. It then forwards the call request to the mobile device through the BS. When the mobile device answers the call, the path is setup. Then voice/data transfer can take place.

Now let us look at what happens when a user roams to another network under the control of another MSC. The MSC of the visited network, on receiving this information from the BS, will make an entry in its VLR, and send this information to the MSC of the home network. The home MSC will make an entry in its HLR about the visited location.

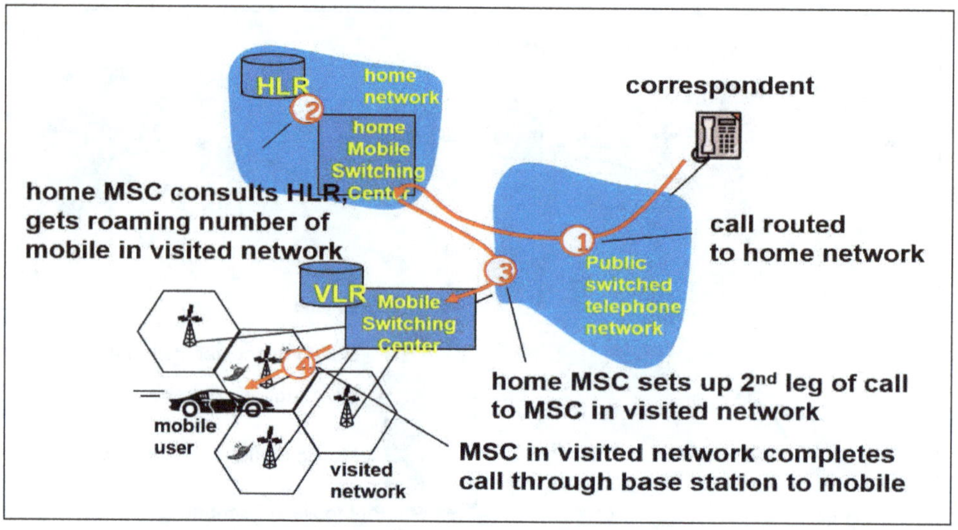


Figure: Call routing to roaming user.

Now, when a caller calls this user, the public switched network will first route the call to the home MSC of this device. Next, the MSC consults its HLR to locate the user being called. The current location pointed to will be that of the visiting network. The call will now be sent to the MSC of the visiting network, which will check its VLR, obtain the details of the BS to which the user is connected, and complete the call.

## Handling Mobility – Call Handoff

When the user is moving from one BSS to another as he is talking, another important aspect to be handled is that his dissociation from one BSS and association with another BSS has to take place without the call being broken or dropped. This is referred to as call handoff. The goal of handoff is to route the call via new the base station without interruption. Handoff can be performed for many reasons. Normally, it is done when there is a stronger signal to/from a new BSS. Such a handoff will help in having continued connectivity, but with less battery drain. It is also done for load balancing; i.e., to free up channels in current BSS. GSM does not mandate why to perform handoff (i.e., the policy), but only the mechanism to be used. The handoff is initiated by the old BS.

There are two cases that we need to consider here: One is a handoff between 2 BSSs with common MSC, and another is a handoff between MSCs.

## Handoff with Common MSC

The steps involved are as follows:

- Old BSS informs MSC of impending handoff, provides list of 1+ new BSSs.
- MSC sets up path (allocates resources) to new BSS.
- New BSS allocates radio channel for use by mobile.
- New BSS signals MSC, and old BSS that it is ready.
- Old BSS tells mobile to perform handoff to new BSS.
- Mobile, and new BSS exchange signal to activate new channel.
- Mobile signals via new BSS to MSC that the handoff is complete. MSC reroutes call.
- MSC's old-BSS resources are released.

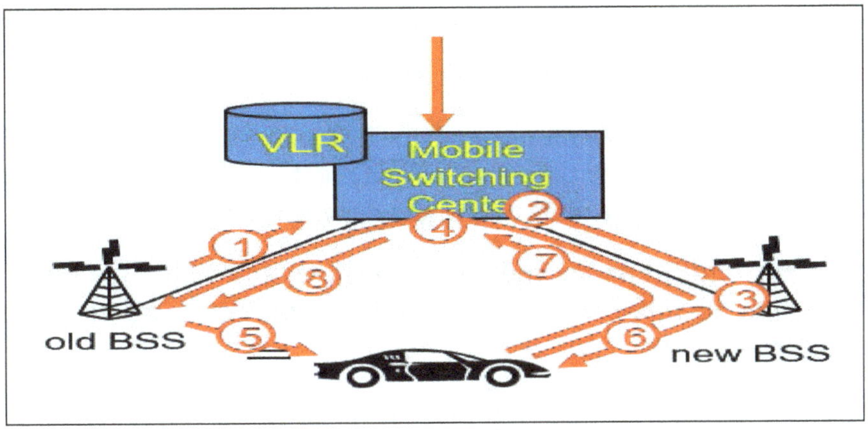


Figure: Handoff within MSC.

## Handoff Between MSCs

When the handoff has to be done across MSCs, the first MSC visited during the call remains as the "anchor" MSC. This anchor keeps track of the next position that the user has moved to. New MSCs add on to the end of the MSC chain as the mobile moves to new MSCs. Call routing in this case will be performed as shown in Fig. If the user moves to yet another MSC's region, that information will be available at the predecessors MSC. Some optimization can be done by feeding the current information to the anchor MSC as well, to reduce the call routing time.

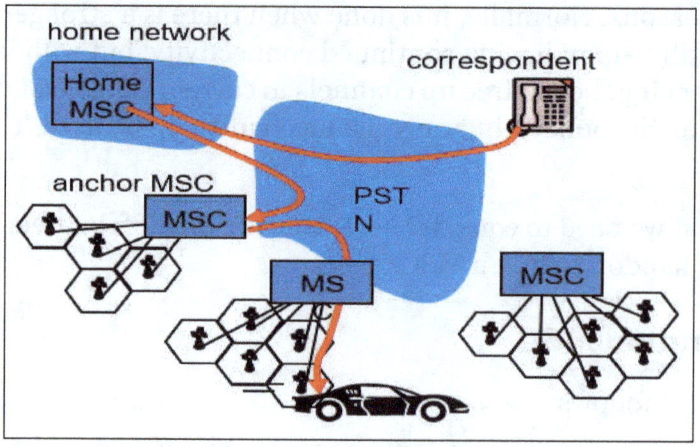


Figure: Call routing in Handoff between MSCs.

## 3G Network

3G is the next generation of wireless network technology that provides high speed bandwidth (high data transfer rates) to handheld devices. The high data transfer rates will allow 3G networks to offer multimedia services combining voice and data. Specifically, 3G wireless networks support the following maximum data transfer rates:

- 2.05 Mbits/second to stationary devices.
- 384 Kbits/second for slowly moving devices, such as a handset carried by a walking user.
- 128 Kbits/second for fast moving devices, such as handsets in moving vehicles.

These data rates are the absolute maximum numbers. For example, in the stationary case, the 2.05 Mb/second rate is for one user hogging the entire capacity of the base station. This data rate will be far lower if there is voice traffic (the actual data rate would depend upon the number of calls in progress).

The maximum data rate of 128Kbits/second for moving devices is about ten times faster than that available with the current 2G wireless networks. Unlike 3G networks, 2G networks were designed to carry voice but not data.

3G wireless networks have the bandwidth to provide converged voice and data services. 3G services will seamlessly combine superior voice quality telephony, high-speed mobile IP services, information technology, rich media, and offer diverse content.

Some characteristics of 3G services that have been proposed are:

- Always-on connectivity. 3G networks use IP connectivity, which is packet based.
- Multi-media services with streaming audio and video.
- Email with full-fledged attachments such as PowerPoint files.
- Instant messaging with video/audio clips.

- Fast downloads of large files such as faxes and PowerPoint files.
- Access to corporate applications.

## Is there A 3G Standard?

The International Telecommunication Union (ITU) is responsible for standardizing 3G. After trying to establish a single 3G standard, ITU finally approved a family of five 3G standards, which are part of the 3G framework known as IMT-2000:

- Three standards based on CDMA, namely CDMA2000, WCDMA, and TD-SCDMA.
- Two standards based on based on TDMA, namely, FDMA/TDMA and TDMA-SC (EDGE).

The CDMA standards are the leading 3G standards. Europe, Japan, and Asia have agreed upon a 3G standard called the Universal Mobile Telecommunications System (UMTS), which is WCDMA operating at 2.1GHz. Note that UMTS and WCDMA are often used as synonyms. In the USA and other parts of Americas, WCDMA will have to use another part of the radio spectrum. Incidentally, most of the world's wireless operators have chosen to use UMTS.

## What are the Advantages of 3G?

3G networks offer users advantages such as:

- New radio spectrum to relieve overcrowding in existing systems.
- More bandwidth, security, and reliability.
- Interoperability between service providers.
- Fixed and variable data rates.
- Asymmetric data rates.
- Backward compatibility of devices with existing networks.
- Always-online devices. 3G will use IP connectivity, IP is packet based (not circuit based).
- Rich multimedia services.

## What are some Disadvantages of 3G?

There are some issues in deploying 3G:

- The cost of upgrading base stations and cellular infrastructure to 3G is very high.
- Requires different handsets and there is the issue of handset availability. 3G handsets will be a complex product. Roaming and making both data/voice work has not yet been demonstrated. Also the higher power requirements (more bits with the same energy/bit) demand a larger handset, shorter talk time, and larger batteries).
- Base stations need to be closer to each other (more cost).
- Tremendous spectrum-license costs, network deployment costs, handset subsidies to subscribers, etc.

- Wireless service providers in Germany and Britain who won spectrum licenses in auctions, paid astronomical prices for them. As a result, they have little money left for building the infrastructure. Consequently, deployment of 3G in Germany and Britain will be delayed.

### What Applications will 3G Enable?

3G represents a paradigm shift from the voice centric world of the previous generations of wireless networks to the multi-media centric world of 3G. Reflecting the high 3G bandwidth and the fact that it is packet based, 3G devices will offer capabilities that are a combination of a phone, PC, and a TV. Examples of services that will be 3G networks can offer are:

- Always-on connection with users paying only when sending or receiving packets.
- Web surfing.
- Instant messaging and email with multimedia attachments.
- Location based services.
- Personalized services, where content can be pushed to users.
- Broadband multimedia data services like video conferencing and streaming video.
- Receiving faxes.
- Global roaming capability.
- Getting maps and directions with a multi-modal user interface.
- Customized entertainment.
- Simultaneous access to multiple services, each service offering some combination of voice, video, data, etc.

### Potential Killer Applications

The high bandwidth of 3G networks will lead to the creation of new services, some of which we have no idea about at this time. The big question is what services will be big revenue makers for the wireless service providers. In 2G networks, the big winners have been short text messaging in GSM networks (Europe and countries other than USA) and image downloads and forwarding on iMode networks in Japan. Two candidate services for big winners in 3G networks are:

- Video conferencing.
- Video messaging.

## 4G Network

4G is the fourth generation of mobile phone technology. It follows on from 3G (third generation) and 2G (second generation) mobile technology. 2G technology launched in the 1990s and made it possible to make digital phone calls and send texts. Then 3G came along in 2003 and made it possible to browse web pages, make video calls and download music and video on the move. 4G technology builds upon what 3G offers but does everything at a much faster speed. Of course, there's now 5G too, which follows the same pattern. It is the fifth generation and it is faster still.

## What are the Benefits of 4G?

The benefits of 4G fall firmly into three categories, which are improved speeds, reduced latency, and crystal-clear voice calls.

## Faster Speeds

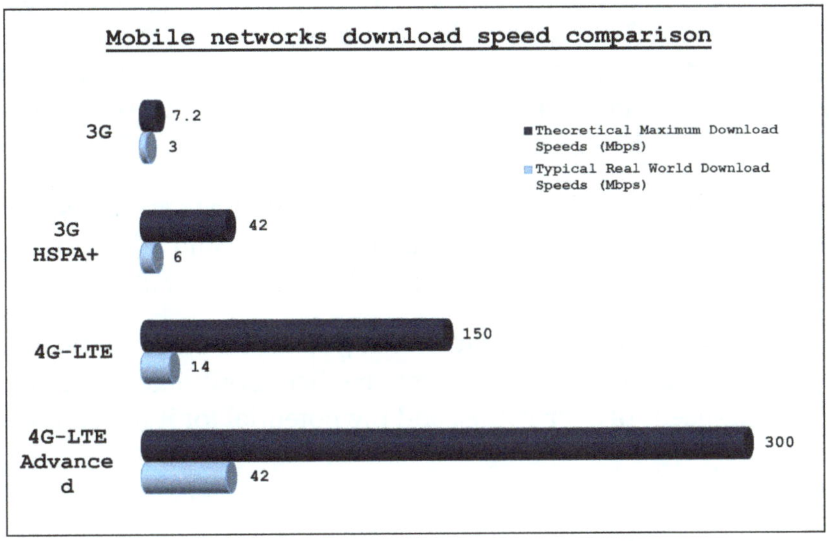


Standard 4G (or 4G LTE) is around five to seven times faster than 3G, offering theoretical speeds of up to around 150Mbps. That equates to maximum potential speeds of around 80Mbps in the real world.

With standard 4G you can download a 2GB HD film in 3 minutes 20 seconds, while it would take over 25 minutes on a standard 3G network.

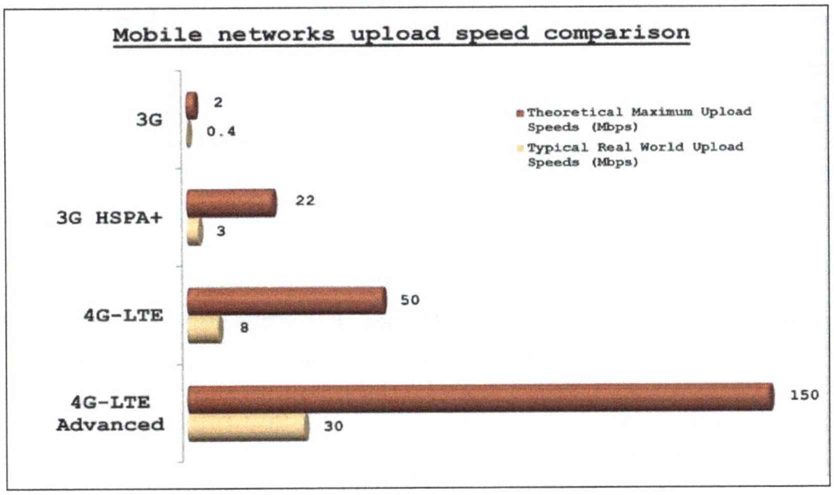

However, a faster version of 4G is also available in many parts of the UK, called 4G LTE-Advanced (also known as LTE-A, 4.5G or 4G+). This offers theoretical speeds of up to 1.5Gbps, but the current crop of LTE-A networks has a maximum potential speed of 300Mbps with real world speeds falling a lot lower.

EE has brought LTE-A to several UK cities, including London, Birmingham, Liverpool, and Manchester, among others. It is not the only network with LTE-A though, as Vodafone has also brought it to London, Birmingham, Manchester, Liverpool, Cardiff, Nottingham, Bristol, and numerous other locations.

Most recently, three has launched a 4G+ service of its own, promising average speeds that are 15% to 33% higher where available. The network has not provided a full list of 4G+ locations but has said that it's been launched in high-demand cities such as London, Birmingham, Cardiff, Edinburgh, and Glasgow. You will know it's working where you are if you see a '4G+' symbol on your phone.

You can expect the availability of LTE-A to increase in the coming years and speeds could get faster still. For example, Ericsson has developed technology which could allow for peak real-world 4G speeds of 1Gbps, though this will not likely be available for a while if at all.

But of course, the main speed boosts will come from 5G, which the UK's main networks are now rolling out. This at the time of writing offers average download speeds in the region of 130-250Mbps, with peak speeds of over 1Gbps, and the potential for it to get far faster still.

### Lower Latency

Download speeds aren't the only thing that has been improved, because 4G also has a better response time than 3G – due to lower 'latency'. This means that a device connected to a 4G mobile network will get a quicker response to a request than the same device connected to a 3G mobile network.

The improved latency times, reduced from 80 milliseconds (3G) to around 45 milliseconds (4G), may not seem that significant on paper. However, they can make a significant difference when playing online games and streaming live video. Find out more about the benefits of 4G for gaming.

On 5G, latency is lowered further still; coming in at around 21-26ms based on Ookla data from late 2019, and likely to further decrease as 5G infrastructure improves.

### Crystal Clear Voice Calls

Voice over LTE (VoLTE) is similar to Voice over Internet Protocol (VoIP), which uses voice apps such as Skype to support voice calls over the internet. Effectively, VoLTE rides on the back of the 4G network and brings crystal clear voice calls and video chat to your 4G mobile phone.

Another benefit is that it allows you to call and text when you only have a 4G connection, where previously you wouldn't have been able to.

## 5G Network

5G is next generation wireless network technology that's expected to change the way people live and work. It will be faster and able to handle more connected devices than the existing 4G LTE network, improvements that will enable a wave of new kinds of tech products. 5G networks began rolling out in the United States and around the world in 2018 and are still in their early days, but experts say the potential is huge.

### Why 5G?

Companies are racing to have the fastest or largest 5G networks. And countries are competing to be the first to deploy fully functional, nationwide 5G. That's because the benefits of the new technology are expected to fuel transformative new technologies, not just for consumers but also for businesses, infrastructure and defense applications.

### Benefits

Much of the hype around 5G has to do with speed. But there are other perks, too. 5G will have greater bandwidth, meaning it can handle many more connected devices than previous networks. That means no more spotty service when you're in a crowded area. And it will enable even more connected devices like smart toothbrushes and self-driving cars.

5G will also reduce latency — the time it takes for a cell phone (or other connected device) to make a request from a server and get a response — to virtually zero. And it will make communication with cloud platforms (think Amazon Web Services and Microsoft Azure) faster and easier.

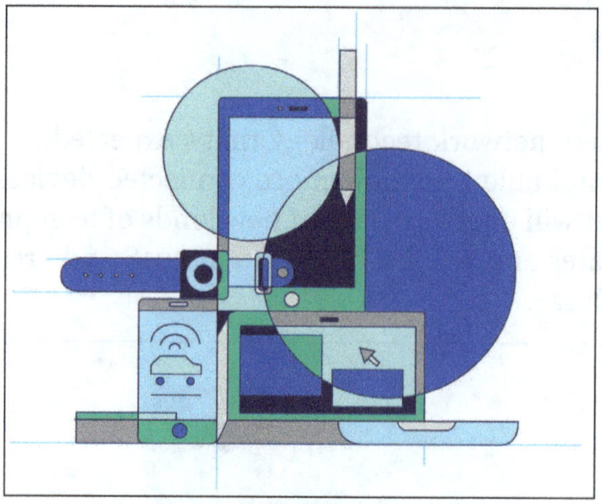

## Working

With 5G, signals run over new radio frequencies, which require updating radios and other equipment on cell towers. There are three different methods for building a 5G network, depending on the type of assets a wireless carrier has: low-band network (wide coverage area but only about 20% faster than 4G), high-band network (superfast speeds but signals don't travel well and struggle to move through hard surfaces) and mid-band network (balances speed and coverage).

Carriers building superfast 5G networks must install tons of small cell sites — about the size of pizza boxes — to light poles, walls or towers, often in relatively small proximity to one another. For that reason, superfast networks are mostly being deployed city by city. Eventually, most US carriers will have a mix of the different network types that will enable both broad coverage and fast speeds.

## Fastness

The fastest 5G networks are expected to be at least 10 times faster than 4G LTE, according to wireless industry trade group GSMA. Some experts say they could eventually be 100 times faster. That's fast enough to download a two hour movie in fewer than 10 seconds, versus around 7 minutes with 4G. Actual download speeds will depend on a number of factors, including location and network traffic.

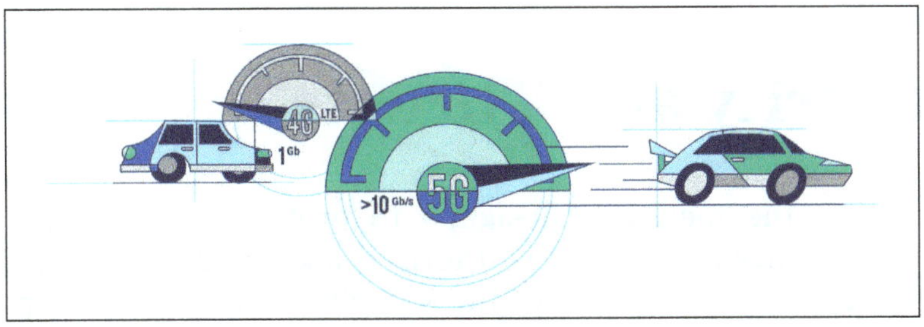


## Usage

In order to connect to and get the benefits of a 5G network, consumers have to have 5G-enabled devices. Samsung, Motorola, Huawei, LG, OnePlus and several other device makers have released 5G phones. Apple is widely expected to release a 5G iPhone later in fall 2020. Some companies – including manufacturers and the NFL – are also working with carriers to install personal 5G networks so they can reap the benefits without waiting for the nationwide rollout.

## Drawbacks

Significant adoption of 5G is going to take years – industry trade group GSMA estimates that by 2025, around half of mobile connections will be 5G (the rest will be older tech, like 4G and 3G). There are also concerns among regulators and others about the security of 5G, especially since crucial technologies such as self-driving cars and healthcare systems will be built on top of the network.

## References


- Cellular-network, computer-science: sciencedirect.com, Retrieved 15, January 2020
- What-is-2g-network: safetouch.com, Retrieved 27, May 2020
- GSM-technology-architecture-its-applications: elprocus.com, Retrieved 09, February 2020
- What-is-4g: 4g.co.uk, Retrieved 22, April 2020
- What-is-5g, business, interactive: edition.cnn.com, Retrieved 22, April 2020

6

# Mobile Security and Threats

Mobile security refers to the measures designed to protect the information stored on and transmitted by computing devices from unauthorized access and hacking. With the usage of computing devices concerns have risen over the security afforded by computing devices. This chapter focuses on mobile security and its related concerns at length for the benefit of the reader.

## Mobile Security

Mobile security or mobile phone security has become increasingly important in mobile computing. Of particular concern is the security of personal and business information now stored on smartphones.

More and more users and businesses employ smartphones as communication tools, but also as a means of planning and organizing their work and private life. Within companies, these technologies are causing profound changes in the organization of information systems and therefore they have become the source of new risks. Indeed, smartphones collect and compile an increasing amount of sensitive information to which access must be controlled to protect the privacy of the user and the intellectual property of the company.

All smartphones, as computers, are preferred targets of attacks. These attacks exploit weaknesses related to smartphones that can come from means of communication like Short Message Service (SMS, aka text messaging), Multimedia Messaging Service (MMS), Wi-Fi networks, Bluetooth and GSM, the de facto global standard for mobile communications. There are also attacks that exploit software vulnerabilities from both the web browser and operating system. Finally, there are forms of malicious software that rely on the weak knowledge of average users.

Different security counter-measures are being developed and applied to smartphones, from security in different layers of software to the dissemination of information to end users. There are good practices to be observed at all levels, from design to use, through the development of operating systems, software layers, and downloadable apps.

### Challenges of Mobile Security

#### Threats

A smartphone user is exposed to various threats when they use their phone. In just the last two quarters of 2012, the number of unique mobile threats grew by 261%, according to ABI Research. These threats can disrupt the operation of the smartphone, and transmit or modify user data. For these reasons, the applications deployed there must guarantee privacy and integrity of the information they handle. In addition, since some apps could themselves be malware, their functionality and activities should be limited (for example, restricting the apps from accessing location information via GPS, blocking access to the user's address book, preventing the transmission of data on the network, sending SMS messages that are billed to the user, etc.).

There are three prime targets for attackers:

- Data: smartphones are devices for data management, therefore they may contain sensitive data like credit card numbers, authentication information, private information, activity logs (calendar, call logs).
- Identity: smartphones are highly customizable, so the device or its contents are associated with a specific person. For example, every mobile device can transmit information related to the owner of the mobile phone contract, and an attacker may want to steal the identity of the owner of a smartphone to commit other offenses.
- Availability: by attacking a smartphone one can limit access to it and deprive the owner of the service.

The source of these attacks are the same actors found in the non-mobile computing space:

- Professionals, whether commercial or military, who focus on the three targets mentioned above. They steal sensitive data from the general public, as well as undertake industrial espionage. They will also use the identity of those attacked to achieve other attacks.
- Thieves who want to gain income through data or identities they have stolen. The thieves will attack many people to increase their potential income.
- Black hat hackers who specifically attack availability. Their goal is to develop viruses, and cause damage to the device. In some cases, hackers have an interest in stealing data on devices.
- Grey hat hackers who reveal vulnerabilities. Their goal is to expose vulnerabilities of the device. Grey hat hackers do not intend on damaging the device or stealing data.

## Consequences

When a smartphone is infected by an attacker, the attacker can attempt several things:

- The attacker can manipulate the smartphone as a zombie machine, that is to say, a machine with which the attacker can communicate and send commands which will be used to send unsolicited messages (spam) via sms or email.
- The attacker can easily force the smartphone to make phone calls. For example, one can use the API (library that contains the basic functions not present in the smartphone) PhoneMakeCall by Microsoft, which collects telephone numbers from any source such as yellow pages, and then call them. But the attacker can also use this method to call paid services, resulting in a charge to the owner of the smartphone. It is also very dangerous because the smartphone could call emergency services and thus disrupt those services.
- A compromised smartphone can record conversations between the user and others and

send them to a third party. This can cause user privacy and industrial security problems.

- An attacker can also steal a user's identity, usurp their identity (with a copy of the user's sim card or even the telephone itself), and thus impersonate the owner. This raises security concerns in countries where smartphones can be used to place orders, view bank accounts or are used as an identity card.
- The attacker can reduce the utility of the smartphone, by discharging the battery. For example, they can launch an application that will run continuously on the smartphone processor, requiring a lot of energy and draining the battery. One factor that distinguishes mobile computing from traditional desktop PCs is their limited performance. Frank Stajano and Ross Anderson first described this form of attack, calling it an attack of "battery exhaustion" or "sleep deprivation torture".
- The attacker can prevent the operation and/or starting of the smartphone by making it unusable. This attack can either delete the boot scripts, resulting in a phone without a functioning OS, or modify certain files to make it unusable (e.g. a script that launches at startup that forces the smartphone to restart) or even embed a startup application that would empty the battery.
- The attacker can remove the personal (photos, music, videos, etc.) or professional data of the user.

## Attacks based on Communication

### Attack based on SMS and MMS

Some attacks derive from flaws in the management of SMS and MMS.

Some mobile phone models have problems in managing binary SMS messages. It is possible, by sending an ill-formed block, to cause the phone to restart, leading to denial of service attacks. If a user with a Siemens S55 received a text message containing a Chinese character, it would lead to a denial of service. In another case, while the standard requires that the maximum size of a Nokia Mail address is 32 characters, some Nokia phones did not verify this standard, so if a user enters an email address over 32 characters, that leads to complete dysfunction of the e-mail handler and puts it out of commission. This attack is called "curse of silence". A study on the safety of the SMS infrastructure revealed that SMS messages sent from the Internet can be used to perform a distributed denial of service (DDoS) attack against the mobile telecommunications infrastructure of a big city. The attack exploits the delays in the delivery of messages to overload the network.

Another potential attack could begin with a phone that sends an MMS to other phones, with an attachment. This attachment is infected with a virus. Upon receipt of the MMS, the user can choose to open the attachment. If it is opened, the phone is infected, and the virus sends an MMS with an infected attachment to all the contacts in the address book. There is a real-world example of this attack: the virus Commwarrior uses the address book and sends MMS messages including an infected file to recipients. A user installs the software, as received via MMS message. Then, the virus began to send messages to recipients taken from the address book.

## Attacks based on Communication

## Attacks based on the GSM Networks

The attacker may try to break the encryption of the mobile network. The GSM network encryption algorithms belong to the family of algorithms called A5. Due to the policy of security through obscurity it has not been possible to openly test the robustness of these algorithms. There were originally two variants of the algorithm: A5/1 and A5/2 (stream ciphers), where the former was designed to be relatively strong, and the latter was designed to be weak on purpose to allow easy cryptanalysis and eavesdropping. ETSI forced some countries (typically outside Europe) to use A5/2. Since the encryption algorithm was made public, it was proved it was possible to break the encryption: A5/2 could be broken on the fly, and A5/1 in about 6 hours . In July 2007, the 3GPP approved a change request to prohibit the implementation of A5/2 in any new mobile phones, which means that is has been decommissioned and is no longer implemented in mobile phones. Stronger public algorithms have been added to the GSM standard, the A5/3 and A5/4 (Block ciphers), otherwise known as KASUMI or UEA1 published by the ETSI. If the network does not support A5/1, or any other A5 algorithm implemented by the phone, then the base station can specify A5/0 which is the null-algorithm, whereby the radio traffic is sent unencrypted. Even in case mobile phones are able to use 3G or 4G which have much stronger encryption than 2G GSM, the base station can downgrade the radio communication to 2G GSM and specify A5/0 (no encryption) . This is the basis for eavesdropping attacks on mobile radio networks using a fake base station commonly called an IMSI catcher.

In addition, tracing of mobile terminals is difficult since each time the mobile terminal is accessing or being accessed by the network, a new temporary identity (TMSI) is allocated to the mobile terminal. The TSMI is used as identity of the mobile terminal the next time it accesses the network. The TMSI is sent to the mobile terminal in encrypted messages.

Once the encryption algorithm of GSM is broken, the attacker can intercept all unencrypted communications made by the victim's smartphone.

## Attacks based on Wi-Fi

An attacker can try to eavesdrop on Wi-Fi communications to derive information (e.g. username, password). This type of attack is not unique to smartphones, but they are very vulnerable to these attacks because very often the Wi-Fi is the only means of communication they have to access the internet. The security of wireless networks (WLAN) is thus an important subject. Initially wireless networks were secured by WEP keys. The weakness of WEP is a short encryption key which is the same for all connected clients. In addition, several reductions in the search space of the keys have been found by researchers. Now, most wireless networks are protected by the WPA security protocol. WPA is based on the "Temporal Key Integrity Protocol (TKIP)" which was designed to allow migration from WEP to WPA on the equipment already deployed. The major improvements in security are the dynamic encryption keys.

For small networks, the WPA is a "pre-shared key" which is based on a shared key. Encryption can be vulnerable if the length of the shared key is short. With limited opportunities for input (i.e. only the numeric keypad) mobile phone users might define short encryption keys that contain only numbers.

This increases the likelihood that an attacker succeeds with a brute-force attack. The successor to WPA, called WPA2, is supposed to be safe enough to withstand a brute force attack.

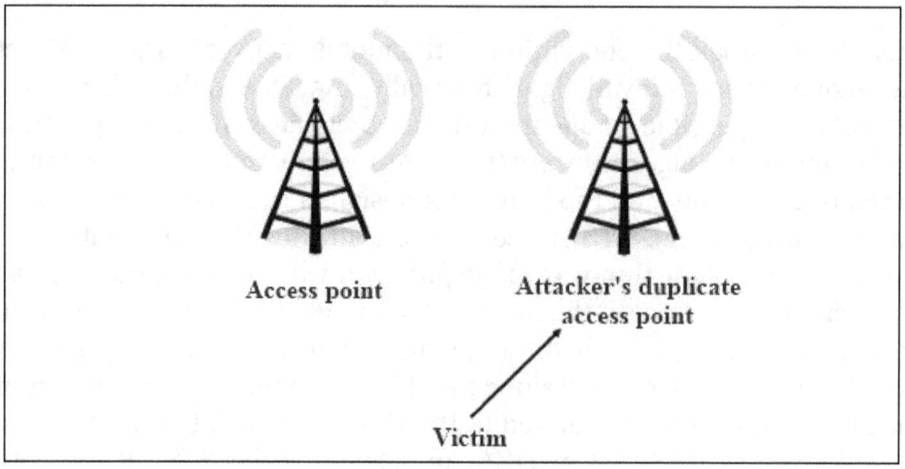


Access Point spoofing

As with GSM, if the attacker succeeds in breaking the identification key, it will be possible to attack not only the phone but also the entire network it is connected to.

Many smartphones for wireless LANs remember they are already connected, and this mechanism prevents the user from having to re-identify with each connection. However, an attacker could create a WIFI access point twin with the same parameters and characteristics as the real network. Using the fact that some smartphones remember the networks, they could confuse the two networks and connect to the network of the attacker who can intercept data if it does not transmit its data in encrypted form.

Lasco is a worm that initially infects a remote device using the SIS file format. SIS file format (Software Installation Script) is a script file that can be executed by the system without user interaction. The smartphone thus believes the file to come from a trusted source and downloads it, infecting the machine.

## Principle of Bluetooth-based Attacks

Security issues related to Bluetooth on mobile devices have been studied and have shown numerous problems on different phones. One easy to exploit vulnerability: unregistered services do not require authentication, and vulnerable applications have a virtual serial port used to control the phone. An attacker only needed to connect to the port to take full control of the device. Another example: a phone must be within reach and Bluetooth in discovery mode. The attacker sends a file via Bluetooth. If the recipient accepts, a virus is transmitted. For example: Cabir is a worm that spreads via Bluetooth connection. The worm searches for nearby phones with Bluetooth in discoverable mode and sends itself to the target device. The user must accept the incoming file and install the program. After installing, the worm infects the machine.

## Attacks based on Vulnerabilities in Software Applications

Other attacks are based on flaws in the OS or applications on the phone.

## Web Browser

The mobile web browser is an emerging attack vector for mobile devices. Just as common Web browsers, mobile web browsers are extended from pure web navigation with widgets and plug-ins, or are completely native mobile browsers.

Jailbreaking the iPhone with firmware 1.1.1 was based entirely on vulnerabilities on the web browser. As a result, the exploitation of the vulnerability described here underlines the importance of the Web browser as an attack vector for mobile devices. In this case, there was a vulnerability based on a stack-based buffer overflow in a library used by the web browser (Libtiff).

A vulnerability in the web browser for Android was discovered in October 2008. As the iPhone vulnerability above, it was due to an obsolete and vulnerable library. A significant difference with the iPhone vulnerability was Android's sandboxing architecture which limited the effects of this vulnerability to the Web browser process.

Smartphones are also victims of classic piracy related to the web: phishing, malicious websites, etc. The big difference is that smartphones do not yet have strong antivirus software available.

## Operating System

Sometimes it is possible to overcome the security safeguards by modifying the operating system itself. As real-world examples, this section covers the manipulation of firmware and malicious signature certificates. These attacks are difficult.

In 2004, vulnerabilities in virtual machines running on certain devices were revealed. It was possible to bypass the bytecode verifier and access the native underlying operating system. The results of this research were not published in detail. The firmware security of Nokia's Symbian Platform Security Architecture (PSA) is based on a central configuration file called SWIPolicy. In 2008 it was possible to manipulate the Nokia firmware before it is installed, and in fact in some downloadable versions of it, this file was human readable, so it was possible to modify and change the image of the firmware. This vulnerability has been solved by an update from Nokia.

In theory smartphones have an advantage over hard drives since the OS files are in ROM, and cannot be changed by malware. However, in some systems it was possible to circumvent this: in the Symbian OS it was possible to overwrite a file with a file of the same name. On the Windows OS, it was possible to change a pointer from a general configuration file to an editable file.

When an application is installed, the signing of this application is verified by a series of certificates. One can create a valid signature without using a valid certificate and add it to the list. In the Symbian OS all certificates are in the directory. With firmware changes explained above it is very easy to insert a seemingly valid but malicious certificate.

## Attacks based on Hardware Vulnerabilities

### Electromagnetic Waveforms

In 2015, researchers at the French government agency ANSSI demonstrated the capability to trigger the voice interface of certain smartphones remotely by using "specific electromagnetic waveforms". The exploit took advantage of antenna-properties of headphone wires while plugged into the audio-output jacks of the vulnerable smartphones and effectively spoofed audio input to inject commands via the audio interface.

### Juice Jacking

Juice Jacking is a method of physical or a hardware vulnerability specific to mobile platforms. Utilizing the dual purpose of the USB charge port, many devices have been susceptible to having data ex-filtrated from, or malware installed on to a mobile device by utilizing malicious charging kiosks set up in public places, or hidden in normal charge adapters.

### Password Cracking

In 2010, researcher from the University of Pennsylvania investigated the possibility of cracking a device's password through a smudge attack (literally imaging the finger smudges on the screen to discern the user's password). The researchers were able to discern the device password up to 68% of the time under certain conditions. Outsiders may perform over-the-shoulder on victims, such as watching specific keystrokes or pattern gestures, to unlock device password or passcode.

### Malicious Software (Malware)

As smartphones are a permanent point of access to the internet (mostly on), they can be compromised as easily as computers with malware. A malware is a computer program that aims to harm the system in which it resides. Trojans, worms and viruses are all considered malware. A Trojan is a program that is on the smartphone and allows external users to connect discreetly. A worm is a program that reproduces on multiple computers across a network. A virus is malicious software designed to spread to other computers by inserting itself into legitimate programs and running programs in parallel. However, it must be said that the malware are far less numerous and important to smartphones as they are to computers.

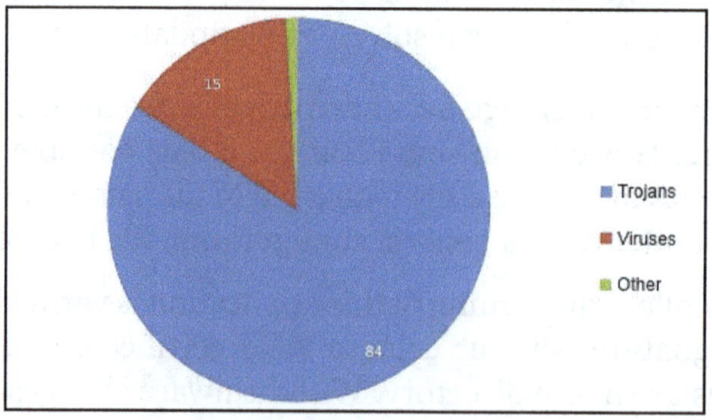


Nonetheless, recent studies show that the evolution of malware in smartphones have rocketed in the last few years posing a threat to analysis and detection.

### The Three Phases of Malware Attacks

Typically an attack on a smartphone made by malware takes place in 3 phases: the infection of a host, the accomplishment of its goal, and the spread of the malware to other systems. Malware often use the resources offered by the infected smartphones. It will use the output devices such as Bluetooth or infrared, but it may also use the address book or email address of the person to infect the user's acquaintances. The malware exploits the trust that is given to data sent by an acquaintance.

## Infection

Infection is the means used by the malware to get into the smartphone, it can either use one of the faults previously presented or may use the gullibility of the user. Infections are classified into four classes according to their degree of user interaction:

Explicit permission:

> The most benign interaction is to ask the user if it is allowed to infect the machine, clearly indicating its potential malicious behavior. This is typical behavior of a proof of concept malware.

Implied permission:

> This infection is based on the fact that the user has a habit of installing software. Most trojans try to seduce the user into installing attractive applications (games, useful applications etc.) that actually contain malware.

Common interaction:

> This infection is related to a common behavior, such as opening an MMS or email.

No interaction:

> The last class of infection is the most dangerous. Indeed, a worm that could infect a smartphone and could infect other smartphones without any interaction would be catastrophic.

## Accomplishment of its Goal

Once the malware has infected a phone it will also seek to accomplish its goal, which is usually one of the following: monetary damage, damage data and/or device, and concealed damage:

M
Monetary damages:

> The attacker can steal user data and either sell them to the same user, or sell to a third party.

Damage:

> Malware can partially damage the device, or delete or modify data on the device.

Concealed damage:

> The two aforementioned types of damage are detectable, but the malware can also leave a backdoor for future attacks or even conduct wiretaps.

## Spread to other Systems

Once the malware has infected a smartphone, it always aims to spread one way or another:

- It can spread through proximate devices using Wi-Fi, Bluetooth and infrared.

- It can also spread using remote networks such as telephone calls or SMS or emails.

## Examples of Malware

Here are various malware that exist in the world of smartphones with a short description of each.

## Viruses and Trojans

- Cabir (also known as Caribe, SybmOS/Cabir, Symbian/Cabir and EPOC.cabir) is the name of a computer worm developed in 2004 that is designed to infect mobile phones running Symbian OS. It is believed to be the first computer worm that can infect mobile phones.
- Commwarrior, found March 7, 2005, is the first worm that can infect many machines from MMS. It is sent in the form of an archive file COMMWARRIOR.ZIP that contains a file COMMWARRIOR.SIS. When this file is executed, Commwarrior attempts to connect to nearby devices by Bluetooth or infrared under a random name. It then attempts to send MMS message to the contacts in the smartphone with different header messages for each person, who receive the MMS and often open them without further verification.
- Phage is the first Palm OS virus that was discovered. It transfers to the Palm from a PC via synchronization. It infects all applications that are in the smartphone and it embeds its own code to function without the user and the system detecting it. All that the system will detect is that its usual applications are functioning.
- RedBrowser is a Trojan which is based on java. The Trojan masquerades as a program called "RedBrowser" which allows the user to visit WAP sites without a WAP connection. During application installation, the user sees a request on their phone that the application needs permission to send messages. Therefore, if the user accepts, RedBrowser can send sms to paid call centers. This program uses the smartphone's connection to social networks (Facebook, Twitter, etc.) to get the contact information for the user's acquaintances (provided the required permissions have been given) and will send them messages.
- WinCE.PmCryptic.A is a malicious software on Windows Mobile which aims to earn money for its authors. It uses the infestation of memory cards that are inserted in the smartphone to spread more effectively.
- CardTrap is a virus that is available on different types of smartphone, which aims to deactivate the system and third party applications. It works by replacing the files used to start the smartphone and applications to prevent them from executing. There are different variants of this virus such as Cardtrap.A for SymbOS devices. It also infects the memory card with malware capable of infecting Windows.
- Ghost Push is a malicious software on Android OS which automatically root the android device and installs malicious applications directly to system partition then unroots the device to prevent users from removing the threat by master reset (The threat can be removed only by reflashing). It cripples the system resources, executes quickly, and harder to detect.

## Ransomware

Mobile ransomware is a type of malware that locks users out of their mobile devices in a pay-to-unlock-your-device ploy, it has grown by leaps and bounds as a threat category since 2014. Specific to mobile computing platforms, users are often less security-conscious, particularly as it pertains to scrutinizing applications and web links trusting the native protection capability of the mobile device operating system. Mobile ransomware poses a significant threat to businesses reliant on instant access and availability of their proprietary information and contacts. The likelihood of a traveling businessman paying a ransom to unlock their device is significantly higher since they are at a disadvantage given inconveniences such as timeliness and less likely direct access to IT staff.

## Spyware

- Flexispy is an application that can be considered as a trojan, based on Symbian. The program sends all information received and sent from the smartphone to a Flexispy server. It was originally created to protect children and spy on adulterous spouses.

## Number of Malware

Below is a diagram which loads the different behaviors of smartphone malware in terms of their effects on smartphones:

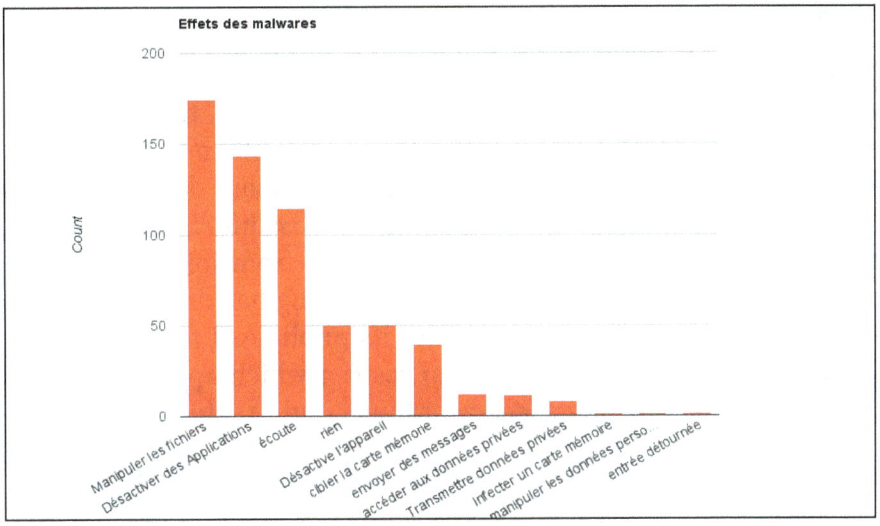


Effects of Malware.

We can see from the graph that at least 50 malwares exhibit no negative behavior, except their ability to spread.

## Portability of Malware across Platforms

There is a multitude of malware. This is partly due to the variety of operating systems on smartphones. However attackers can also choose to make their malware target multiple platforms, and malware can be found which attacks an OS but is able to spread to different systems.

To begin with, malware can use runtime environments like Java virtual machine or the .NET

Framework. They can also use other libraries present in many operating systems. Other malware carry several executable files in order to run in multiple environments and they utilize these during the propagation process. In practice, this type of malware requires a connection between the two operating systems to use as an attack vector. Memory cards can be used for this purpose, or synchronization software can be used to propagate the virus.

## Countermeasures

The security mechanisms in place to counter the threats described above are presented in this section. They are divided into different categories, as all do not act at the same level, and they range from the management of security by the operating system to the behavioral education of the user. The threats prevented by the various measures are not the same depending on the case. Considering the two cases mentioned above, in the first case one would protect the system from corruption by an application, and in the second case the installation of a suspicious software would be prevented.

## Security in Operating Systems

The first layer of security within a smartphone is at the level of the operating system (OS). Beyond the usual roles of an operating system (e.g. resource management, scheduling processes) on a smartphone, it must also establish the protocols for introducing external applications and data without introducing risk.

A central idea found in the mobile operating systems is the idea of a sandbox. Since smartphones are currently being designed to accommodate many applications, they must put in place mechanisms to ensure these facilities are safe for themselves, for other applications and data on the system, and the user. If a malicious program manages to reach a device, it is necessary that the vulnerable area presented by the system be as small as possible. Sandboxing extends this idea to compartmentalize different processes, preventing them from interacting and damaging each other. Based on the history of operating systems, sandboxing has different implementations. For example, where iOS will focus on limiting access to its public API for applications from the App Store by default, Managed Open In allows you to restrict which apps can access which types of data. Android bases its sandboxing on its legacy of Linux and TrustedBSD.

The following points highlight mechanisms implemented in operating systems, especially Android.

Rootkit Detectors:

> The intrusion of a rootkit in the system is a great danger in the same way as on a computer. It is important to prevent such intrusions, and to be able to detect them as often as possible. Indeed, there is concern that with this type of malicious program, the result could be a partial or complete bypass of the device security, and the acquisition of administrator rights by the attacker. If this happens, then nothing prevents the attacker from studying or disabling the safety features that were circumvented, deploying the applications they want, or disseminating a method of intrusion by a rootkit to a wider audience. We can cite, as a defense mechanism, the Chain of trust in iOS. This mechanism relies on the signature of the different applications required to start the operating system, and a certificate signed by

Apple. In the event that the signature checks are inconclusive, the device detects this and stops the boot-up. If the Operating System is compromised due to Jailbreaking, root kit detection may not work if it is disabled by the Jailbreak method or software is loaded after Jailbreak disables Rootkit Detection.

Process isolation:

Android uses mechanisms of user process isolation inherited from Linux. Each application has a user associated with it, and a tuple (UID, GID). This approach serves as a sandbox: while applications can be malicious, they can not get out of the sandbox reserved for them by their identifiers, and thus cannot interfere with the proper functioning of the system. For example, since it is impossible for a process to end the process of another user, an application can thus not stop the execution of another.

File permissions:

From the legacy of Linux, there are also filesystem permissions mechanisms. They help with sandboxing: a process can not edit any files it wants. It is therefore not possible to freely corrupt files necessary for the operation of another application or system. Furthermore, in Android there is the method of locking memory permissions. It is not possible to change the permissions of files installed on the SD card from the phone, and consequently it is impossible to install applications.

Memory Protection:

In the same way as on a computer, memory protection prevents privilege escalation. Indeed, if a process managed to reach the area allocated to other processes, it could write in the memory of a process with rights superior to their own, with root in the worst case, and perform actions which are beyond its permissions on the system. It would suffice to insert function calls are authorized by the privileges of the malicious application.

Development through runtime environments:

Software is often developed in high-level languages, which can control what is being done by a running program. For example, Java Virtual Machines continuously monitor the actions of the execution threads they manage, monitor and assign resources, and prevent malicious actions. Buffer overflows can be prevented by these controls.

## Security Software

Above the operating system security, there is a layer of security software. This layer is composed of individual components to strengthen various vulnerabilities: prevent malware, intrusions, the identification of a user as a human, and user authentication. It contains software components that have learned from their experience with computer security; however, on smartphones, this software must deal with greater constraints.

Antivirus and firewall:

An antivirus software can be deployed on a device to verify that it is not infected by a known threat, usually by signature detection software that detects malicious executable files. A firewall, meanwhile, can watch over the existing traffic on the network and ensure that a malicious application does not seek to communicate through it. It may equally verify that an installed application does not seek to establish suspicious communication, which may prevent an intrusion attempt.

Visual Notifications:

In order to make the user aware of any abnormal actions, such as a call they did not initiate, one can link some functions to a visual notification that is impossible to circumvent. For example, when a call is triggered, the called number should always be displayed. Thus, if a call is triggered by a malicious application, the user can see, and take appropriate action.

Turing test:

In the same vein as above, it is important to confirm certain actions by a user decision. The Turing test is used to distinguish between a human and a virtual user, and it often comes as a captcha.

Biometric identification:

Another method to use is biometrics. Biometrics is a technique of identifying a person by means of their morphology(by recognition of the eye or face, for example) or their behavior (their signature or way of writing for example). One advantage of using biometric security is that users can avoid having to remember a password or other secret combination to authenticate and prevent malicious users from accessing their device. In a system with strong biometric security, only the primary user can access the smartphone.

### Resource Monitoring in the Smartphone

When an application passes the various security barriers, it can take the actions for which it was designed. When such actions are triggered, the activity of a malicious application can be sometimes detected if one monitors the various resources used on the phone. Depending on the goals of the malware, the consequences of infection are not always the same; all malicious applications are not intended to harm the devices on which they are deployed. The following sections describe different ways to detect suspicious activity.

Battery:

Some malware is aimed at exhausting the energy resources of the phone. Monitoring the energy consumption of the phone can be a way to detect certain malware applications.

Memory usage:

Memory usage is inherent in any application. However, if one finds that a substantial proportion of memory is used by an application, it may be flagged as suspicious.

Network traffic:

On a smartphone, many applications are bound to connect via the network, as part of their normal operation. However, an application using a lot of bandwidth can be strongly suspected of attempting to communicate a lot of information, and disseminate data to many other devices. This observation only allows a suspicion, because some legitimate applications can be very resource-intensive in terms of network communications, the best example being streaming video.

Services:

One can monitor the activity of various services of a smartphone. During certain moments, some services should not be active, and if one is detected, the application should be suspected. For example, the sending of an SMS when the user is filming video: this communication does not make sense and is suspicious; malware may attempt to send SMS while its activity is masked.

The various points mentioned above are only indications and do not provide certainty about the legitimacy of the activity of an application. However, these criteria can help target suspicious applications, especially if several criteria are combined.

## Network Surveillance

Network traffic exchanged by phones can be monitored. One can place safeguards in network routing points in order to detect abnormal behavior. As the mobile's use of network protocols is much more constrained than that of a computer, expected network data streams can be predicted (e.g. the protocol for sending an SMS), which permits detection of anomalies in mobile networks.

Spam filters:

As is the case with email exchanges, we can detect a spam campaign through means of mobile communications (SMS, MMS). It is therefore possible to detect and minimize this kind of attempt by filters deployed on network infrastructure that is relaying these messages.

Encryption of stored or transmitted information:

Because it is always possible that data exchanged can be intercepted, communications, or even information storage, can rely on encryption to prevent a malicious entity from using any data obtained during communications. However, this poses the problem of key exchange for encryption algorithms, which requires a secure channel.

Telecom network monitoring:

The networks for SMS and MMS exhibit predictable behavior, and there is not as much liberty compared with what one can do with protocols such as TCP or UDP. This implies that one cannot predict the use made of the common protocols of the web; one might generate very little traffic by consulting simple pages, rarely, or generate heavy traffic by using video streaming. On the other hand, messages exchanged via mobile phone have a framework and a specific model, and the user does not, in a normal case, have the freedom to intervene in the details of these communications. Therefore, if an abnormality is found in the flux of network data in the mobile networks, the potential threat can be quickly detected.

## Manufacturer Surveillance

In the production and distribution chain for mobile devices, it is the responsibility of manufacturers to ensure that devices are delivered in a basic configuration without vulnerabilities. Most users are not experts and many of them are not aware of the existence of security vulnerabilities, so the device configuration as provided by manufacturers will be retained by many users. Below are listed several points which manufacturers should consider.

Remove debug mode:

> Phones are sometimes set in a debug mode during manufacturing, but this mode must be disabled before the phone is sold. This mode allows access to different features, not intended for routine use by a user. Due to the speed of development and production, distractions occur and some devices are sold in debug mode. This kind of deployment exposes mobile devices to exploits that utilize this oversight.

Default settings:

> When a smartphone is sold, its default settings must be correct, and not leave security gaps. The default configuration is not always changed, so a good initial setup is essential for users. There are, for example, default configurations that are vulnerable to denial of service attacks.

Security audit of apps:

> Along with smart phones, appstores have emerged. A user finds themselves facing a huge range of applications. This is especially true for providers who manage appstores because they are tasked with examining the apps provided, from different points of view (e.g. security, content). The security audit should be particularly cautious, because if a fault is not detected, the application can spread very quickly within a few days, and infect a significant number of devices.

Detect suspicious applications demanding rights:

> When installing applications, it is good to warn the user against sets of permissions that, grouped together, seem potentially dangerous, or at least suspicious. Frameworks like such as Kirin, on Android, attempt to detect and prohibit certain sets of permissions.

Revocation procedures:

> Along with appstores appeared a new feature for mobile apps: remote revocation. First developed by Android, this procedure can remotely and globally uninstall an application, on any device that has it. This means the spread of a malicious application that managed to evade security checks can be immediately stopped when the threat is discovered.

Avoid heavily customized systems:

> Manufacturers are tempted to overlay custom layers on existing operating systems, with the dual purpose of offering customized options and disabling or charging for certain features. This has the dual effect of risking the introduction of new bugs in the system, coupled

with an incentive for users to modify the systems to circumvent the manufacturer's restrictions. These systems are rarely as stable and reliable as the original, and may suffer from phishing attempts or other exploits.

Improve software patch processes:

New versions of various software components of a smartphone, including operating systems, are regularly published. They correct many flaws over time. Nevertheless, manufacturers often do not deploy these updates to their devices in a timely fashion, and sometimes not at all. Thus, vulnerabilities persist when they could be corrected, and if they are not, since they are known, they are easily exploitable.

## User Awareness

Much malicious behavior is allowed by the carelessness of the user. From simply not leaving the device without a password, to precise control of permissions granted to applications added to the smartphone, the user has a large responsibility in the cycle of security: to not be the vector of intrusion. This precaution is especially important if the user is an employee of a company that stores business data on the device. Detailed below are some precautions that a user can take to manage security on a smartphone.

A recent survey by internet security experts BullGuard showed a lack of insight into the rising number of malicious threats affecting mobile phones, with 53% of users claiming that they are unaware of security software for Smartphones. A further 21% argued that such protection was unnecessary, and 42% admitted it hadn't crossed their mind ("Using APA," 2011). These statistics show consumers are not concerned about security risks because they believe it is not a serious problem. The key here is to always remember smartphones are effectively handheld computers and are just as vulnerable.

Being skeptical:

A user should not believe everything that may be presented, as some information may be phishing or attempting to distribute a malicious application. It is therefore advisable to check the reputation of the application that they want to buy before actually installing it.

Permissions given to applications:

The mass distribution of applications is accompanied by the establishment of different permissions mechanisms for each operating system. It is necessary to clarify these permissions mechanisms to users, as they differ from one system to another, and are not always easy to understand. In addition, it is rarely possible to modify a set of permissions requested by an application if the number of permissions is too great. But this last point is a source of risk because a user can grant rights to an application, far beyond the rights it needs. For example, a note taking application does not require access to the geolocation service. The user must ensure the privileges required by an application during installation and should not accept the installation if requested rights are inconsistent.

Be careful:

Protection of a user's phone through simple gestures and precautions, such as locking the smartphone when it is not in use, not leaving their device unattended, not trusting applications, not storing sensitive data, or encrypting sensitive data that cannot be separated from the device.

Ensure data:

Smartphones have a significant memory and can carry several gigabytes of data. The user must be careful about what data it carries and whether they should be protected. While it is usually not dramatic if a song is copied, a file containing bank information or business data can be more risky. The user must have the prudence to avoid the transmission of sensitive data on a smartphone, which can be easily stolen. Furthermore, when a user gets rid of a device, they must be sure to remove all personal data first.

These precautions are measures that leave no easy solution to the intrusion of people or malicious applications in a smartphone. If users are careful, many attacks can be defeated, especially phishing and applications seeking only to obtain rights on a device.

## Centralized Storage of Text Messages

One form of mobile protection allows companies to control the delivery and storage of text messages, by hosting the messages on a company server, rather than on the sender or receiver's phone. When certain conditions are met, such as an expiration date, the messages are deleted.

## Limitations of Certain Security Measures

The security mechanisms mentioned in this article are to a large extent inherited from knowledge and experience with computer security. The elements composing the two device types are similar, and there are common measures that can be used, such as antivirus and firewall. However, the implementation of these solutions is not necessarily possible or at least highly constrained within a mobile device. The reason for this difference is the technical resources offered by computers and mobile devices: even though the computing power of smartphones is becoming faster, they have other limitations than their computing power.

- Single-task system: Some operating systems, including some still commonly used, are single-tasking. Only the foreground task is executed. It is difficult to introduce applications such as antivirus and firewall on such systems, because they could not perform their monitoring while the user is operating the device, when there would be most need of such monitoring.
- Energy autonomy: A critical one for the use of a smartphone is energy autonomy. It is important that the security mechanisms not consume battery resources, without which the autonomy of devices will be affected dramatically, undermining the effective use of the smartphone.
- Network Directly related to battery life, network utilization should not be too high. It is indeed one of the most expensive resources, from the point of view of energy consumption. Nonetheless, some calculations may need to be relocated to remote servers in order to preserve the battery. This balance can make implementation of certain intensive computation mechanisms a delicate proposition.

Furthermore, it should be noted that it is common to find that updates exist, or can be developed or deployed, but this is not always done. One can, for example, find a user who does not know that there is a newer version of the operating system compatible with the smartphone, or a user may discover known vulnerabilities that are not corrected until the end of a long development cycle, which allows time to exploit the loopholes.

### Next Generation of Mobile Security

There is expected to be four mobile environments that will make up the security framework:

Rich operating system:

> In this category will fall traditional Mobile OS like Android, iOS, Symbian OS or Windows Phone. They will provide the traditional functionaity and security of an OS to the applications.

Secure Operating System (Secure OS):

> A secure kernel which will run in parallel with a fully featured Rich OS, on the same processor core. It will include drivers for the Rich OS ("normal world") to communicate with the secure kernel ("secure world"). The trusted infrastructure could include interfaces like the display or keypad to regions of PCI-E address space and memories.

Trusted Execution Environment (TEE):

> Made up of hardware and software. It helps in the control of access rights and houses sensitive applications, which need to be isolated from the Rich OS. It effectively acts as a firewall between the "normal world" and "secure world".

Secure Element (SE):

> The SE consists of tamper resistant hardware and associated software. It can provide high levels of security and work in tandem with the TEE. The SE will be mandatory for hosting proximity payment applications or official electronic signatures.

## Malware

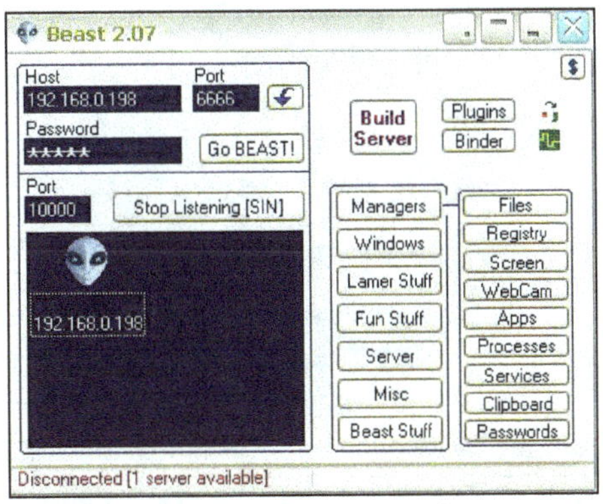


Beast, a Windows-based backdoor Trojan horse.

Malware, short for malicious software, is any software used to disrupt computer operations, gather sensitive information, gain access to private computer systems, or display unwanted advertising. Before the term malware was coined by Yisrael Radai in 1990, malicious software was referred to as computer viruses. The first category of malware propagation concerns parasitic software fragments that attach themselves to some existing executable content. The fragment may be machine code that infects some existing application, utility, or system program, or even the code used to boot a computer system. Malware is defined by its malicious intent, acting against the requirements of the computer user, and does not include software that causes unintentional harm due to some deficiency.

Malware may be stealthy, intended to steal information or spy on computer users for an extended period without their knowledge, as for example Regin, or it may be designed to cause harm, often as sabotage (e.g., Stuxnet), or to extort payment (CryptoLocker). 'Malware' is an umbrella term used to refer to a variety of forms of hostile or intrusive software, including computer viruses, worms, trojan horses, ransomware, spyware, adware, scareware, and other malicious programs. It can take the form of executable code, scripts, active content, and other software. Malware is often disguised as, or embedded in, non-malicious files. As of 2011 the majority of active malware threats were worms or trojans rather than viruses.

In law, malware is sometimes known as a computer contaminant, as in the legal codes of several U.S. states.

Spyware or other malware is sometimes found embedded in programs supplied officially by companies, e.g., downloadable from websites, that appear useful or attractive, but may have, for example, additional hidden tracking functionality that gathers marketing statistics. An example of such software, which was described as illegitimate, is the Sony rootkit, a Trojan embedded into CDs sold by Sony, which silently installed and concealed itself on purchasers' computers with the intention of preventing illicit copying; it also reported on users' listening habits, and unintentionally created vulnerabilities that were exploited by unrelated malware.

Software such as anti-virus, anti-malware, and firewalls are used to protect against activity identified as malicious, and to recover from attacks.

## Purposes

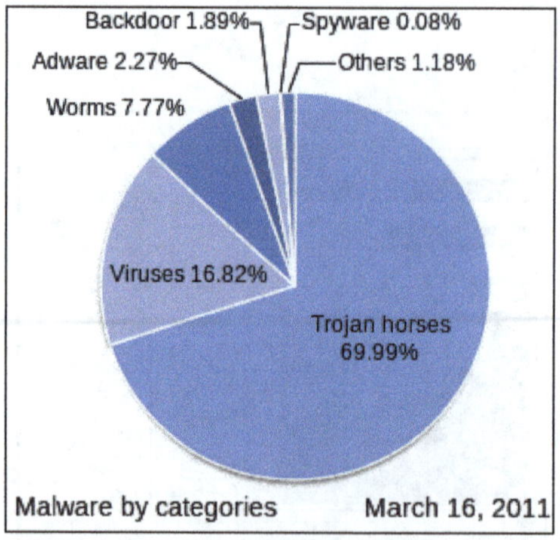


Malware by categories on 16 March 2011.

Many early infectious programs, including the first Internet Worm, were written as experiments or pranks. Today, malware is used by both black hat hackers and governments, to steal personal, financial, or business information.

Malware is sometimes used broadly against government or corporate websites to gather guarded information, or to disrupt their operation in general. However, malware is often used against individuals to gain information such as personal identification numbers or details, bank or credit card numbers, and passwords. Left unguarded, personal and networked computers can be at considerable risk against these threats. (These are most frequently defended against by various types of firewall, anti-virus software, and network hardware).

Since the rise of widespread broadband Internet access, malicious software has more frequently been designed for profit. Since 2003, the majority of widespread viruses and worms have been designed to take control of users' computers for illicit purposes. Infected "zombie computers" are used to send email spam, to host contraband data such as child pornography, or to engage in distributed denial-of-service attacks as a form of extortion.

Programs designed to monitor users' web browsing, display unsolicited advertisements, or redirect affiliate marketing revenues are called spyware. Spyware programs do not spread like viruses; instead they are generally installed by exploiting security holes. They can also be hidden and packaged together with unrelated user-installed software.

Ransomware affects an infected computer in some way, and demands payment to reverse the damage. For example, programs such as CryptoLocker encrypt files securely, and only decrypt them on payment of a substantial sum of money.

Some malware is used to generate money by click fraud, making it appear that the computer user has clicked an advertising link on a site, generating a payment from the advertiser. It was estimated in 2012 that about 60 to 70% of all active malware used some kind of click fraud, and 22% of all ad-clicks were fraudulent.

Malware is usually used for criminal purposes, but can be used for sabotage, often without direct benefit to the perpetrators. One example of sabotage was Stuxnet, used to destroy very specific industrial equipment. There have been politically motivated attacks that have spread over and shut down large computer networks, including massive deletion of files and corruption of master boot records, described as "computer killing". Such attacks were made on Sony Pictures Entertainment (25 November 2014, using malware known as Shamoon or W32.Disttrack) and Saudi Aramco (August 2012).

## Proliferation

Preliminary results from Symantec published in 2008 suggested that "the release rate of malicious code and other unwanted programs may be exceeding that of legitimate software applications." According to F-Secure, "As much malware [was] produced in 2007 as in the previous 20 years altogether." Malware's most common pathway from criminals to users is through the Internet: primarily by e-mail and the World Wide Web.

The prevalence of malware as a vehicle for Internet crime, along with the challenge of anti-malware software to keep up with the continuous stream of new malware, has seen the adoption of a new mindset for individuals and businesses using the Internet. With the amount of malware currently being distributed, some percentage of computers are currently assumed to be infected. For businesses, especially those that sell mainly over the Internet, this means they need to find a way to operate despite security concerns. The result is a greater emphasis on back-office protection designed to protect against advanced malware operating on customers' computers. A 2013 Webroot study shows that 64% of companies allow remote access to servers for 25% to 100% of their workforce and that companies with more than 25% of their employees accessing servers remotely have higher rates of malware threats.

On 29 March 2010, Symantec Corporation named Shaoxing, China, as the world's malware capital. A 2011 study from the University of California, Berkeley, and the Madrid Institute for Advanced Studies published an article in *Software Development Technologies*, examining how entrepreneurial hackers are helping enable the spread of malware by offering access to computers for a price. Microsoft reported in May 2011 that one in every 14 downloads from the Internet may now contain malware code. Social media, and Facebook in particular, are seeing a rise in the number of tactics used to spread malware to computers.

A 2014 study found that malware is being increasingly aimed at mobile devices such as smartphones as they increase in popularity.

### Infectious Malware: Viruses and Worms

The best-known types of malware, viruses and worms, are known for the manner in which they spread, rather than any specific types of behavior. The term *computer virus* is used for a program that embeds itself in some other executable software (including the operating system itself) on the target system without the user's consent and when that is run causes the virus to spread to other executables. On the other hand, a *worm* is a stand-alone malware program that *actively* transmits itself over a network to infect other computers. These definitions lead to the observation that a virus requires the user to run an infected program or operating system for the virus to spread, whereas a worm spreads itself.

### Concealment: Viruses, Trojan Horses, Rootkits, Backdoors and Evasion

These categories are not mutually exclusive, so malware may use multiple techniques. This section only applies to malware designed to operate undetected, not sabotage and ransomware.

### Viruses

A computer program usually hidden within another seemingly innocuous program that produces copies of itself and inserts them into other programs or files, and that usually performs a malicious action (such as destroying data).

### Trojan Horses

In computing, Trojan horse, or Trojan, is any malicious computer program which misrepresents itself to appear useful, routine, or interesting in order to persuade a victim to install it. The term is derived from the Ancient Greek story of the wooden horse that was used to help Greek troops invade the city of Troy by stealth.

Trojans are generally spread by some form of social engineering, for example where a user is duped into executing an e-mail attachment disguised to be unsuspicious, (e.g., a routine form to be filled in), or by drive-by download. Although their payload can be anything, many modern forms act as a backdoor, contacting a controller which can then have unauthorized access to the affected computer. While Trojans and backdoors are not easily detectable by themselves, computers may appear to run slower due to heavy processor or network usage.

Unlike computer viruses and worms, Trojans generally do not attempt to inject themselves into other files or otherwise propagate themselves.

## Rootkits

Once a malicious program is installed on a system, it is essential that it stays concealed, to avoid detection. Software packages known as *rootkits* allow this concealment, by modifying the host's operating system so that the malware is hidden from the user. Rootkits can prevent a malicious process from being visible in the system's list of processes, or keep its files from being read.

Some malicious programs contain routines to defend against removal, not merely to hide themselves. An early example of this behavior is recorded in the Jargon File tale of a pair of programs infesting a Xerox CP-V time sharing system:

> Each ghost-job would detect the fact that the other had been killed, and would start a new copy of the recently stopped program within a few milliseconds. The only way to kill both ghosts was to kill them simultaneously (very difficult) or to deliberately crash the system.

## Backdoors

A backdoor is a method of bypassing normal authentication procedures, usually over a connection to a network such as the Internet. Once a system has been compromised, one or more backdoors may be installed in order to allow access in the future, invisibly to the user.

The idea has often been suggested that computer manufacturers preinstall backdoors on their systems to provide technical support for customers, but this has never been reliably verified. It was reported in 2014 that US government agencies had been diverting computers purchased by those considered "targets" to secret workshops where software or hardware permitting remote access by the agency was installed, considered to be among the most productive operations to obtain access to networks around the world. Backdoors may be installed by Trojan horses, worms, implants, or other methods.

## Evasion

Since the beginning of 2015, a sizable portion of malware utilizes a combination of many techniques designed to avoid detection and analysis:

- The most common evasion technique is when the malware evades analysis and detection by fingerprinting the environment when executed.

- The second most common evasion technique is confusing automated tools' detection methods. This allows malware to avoid detection by technologies such as signature-based antivirus software by changing the server used by the malware.
- The third most common evasion technique is timing-based evasion. This is when malware runs at certain times or following certain actions taken by the user, so it executes during certain vulnerable periods, such as during the boot process, while remaining dormant the rest of the time.
- The fourth most common evasion technique is done by obfuscating internal data so that automated tools do not detect the malware.
- An increasingly common technique is adware that uses stolen certificates to disable anti-malware and virus protection; technical remedies are available to deal with the adware.

Nowadays, one of the most sophisticated and stealthy ways of evasion is to use information hiding techniques, namely stegomalware.

### Vulnerability to Malware

- In this context, and throughout, what is called the "system" under attack may be anything from a single application, through a complete computer and operating system, to a large network.
- Various factors make a system more vulnerable to malware.

### Security Defects in Software

Malware exploits security defects (security bugs or vulnerabilities) in the design of the operating system, in applications (such as browsers, e.g. older versions of Microsoft Internet Explorer supported by Windows XP), or in vulnerable versions of browser plugins such as Adobe Flash Player, Adobe Acrobat or Reader, or Java SE. Sometimes even installing new versions of such plugins does not automatically uninstall old versions. Security advisories from plug-in providers announce security-related updates. Common vulnerabilities are assigned CVE IDs and listed in the US National Vulnerability Database. Secunia PSI is an example of software, free for personal use, that will check a PC for vulnerable out-of-date software, and attempt to update it.

Malware authors target bugs, or loopholes, to exploit. A common method is exploitation of a buffer overrun vulnerability, where software designed to store data in a specified region of memory does not prevent more data than the buffer can accommodate being supplied. Malware may provide data that overflows the buffer, with malicious executable code or data after the end; when this payload is accessed it does what the attacker, not the legitimate software, determines.

### Insecure Ign or User Error

Early PCs had to be booted from floppy disks. When built-in hard drives became common, the operating system was normally started from them, but it was possible to boot from another boot device if available, such as a floppy disk, CD-ROM, DVD-ROM, USB flash drive or network. It was common to configure the computer to boot from one of these devices when available. Normally none would be available; the user would intentionally insert, say, a CD into the optical drive to boot the computer in some special way, for example, to install an operating system. Even without booting, computers can be configured to execute software on some media as soon as they become available, e.g. to autorun a CD or USB device when inserted.

Malicious software distributors would trick the user into booting or running from an infected device or medium. For example, a virus could make an infected computer add autorunnable code to any USB stick plugged into it. Anyone who then attached the stick to another computer set to autorun from USB would in turn become infected, and also pass on the infection in the same way. More generally, any device that plugs into a USB port - even lights, fans, speakers, toys, or peripherals such as a digital microscope - can be used to spread malware. Devices can be infected during manufacturing or supply if quality control is inadequate.

This form of infection can largely be avoided by setting up computers by default to boot from the internal hard drive, if available, and not to autorun from devices. Intentional booting from another device is always possible by pressing certain keys during boot.

Older email software would automatically open HTML email containing potentially malicious JavaScript code. Users may also execute disguised malicious email attachments and infected executable files supplied in other ways.

## Over-privileged users and over-privileged Code

In computing, privilege refers to how much a user or program is allowed to modify a system. In poorly designed computer systems, both users and programs can be assigned more privileges than they should be, and malware can take advantage of this. The two ways that malware does this is through overprivileged users and overprivileged code.

Some systems allow all users to modify their internal structures, and such users today would be considered over-privileged users. This was the standard operating procedure for early microcomputer and home computer systems, where there was no distinction between an *administrator* or *root*, and a regular user of the system. In some systems, non-administrator users are over-privileged by design, in the sense that they are allowed to modify internal structures of the system. In some environments, users are over-privileged because they have been inappropriately granted administrator or equivalent status.

Some systems allow code executed by a user to access all rights of that user, which is known as over-privileged code. This was also standard operating procedure for early microcomputer and home computer systems. Malware, running as over-privileged code, can use this privilege to subvert the system. Almost all currently popular operating systems, and also many scripting applications allow code too many privileges, usually in the sense that when a user executes code, the system allows that code all rights of that user. This makes users vulnerable to malware in the form of e-mail attachments, which may or may not be disguised.

## Use of the Same Operating System

- Homogeneity can be a vulnerability. For example, when all computers in a network run the same operating system, upon exploiting one, one worm can exploit them all: In particular,

Microsoft Windows or Mac OS X have such a large share of the market that an exploited vulnerability concentrating on either operating system could subvert a large number of systems. Introducing diversity purely for the sake of robustness, such as adding Linux computers, could increase short-term costs for training and maintenance. However, as long as all the nodes are not part of the same directory service for authentication, having a few diverse nodes could deter total shutdown of the network and allow those nodes to help with recovery of the infected nodes. Such separate, functional redundancy could avoid the cost of a total shutdown, at the cost of increased complexity and reduced usability in terms of single sign-on authentication.

## Anti-malware Strategies

As malware attacks become more frequent, attention has begun to shift from viruses and spyware protection, to malware protection, and programs that have been specifically developed to combat malware.

## Anti-virus and Anti-malware Software

A specific component of anti-virus and anti-malware software, commonly referred to as an on-access or real-time scanner, hooks deep into the operating system's core or kernel and functions in a manner similar to how certain malware itself would attempt to operate, though with the user's informed permission for protecting the system. Any time the operating system accesses a file, the on-access scanner checks if the file is a 'legitimate' file or not. If the file is identified as malware by the scanner, the access operation will be stopped, the file will be dealt with by the scanner in a pre-defined way (how the anti-virus program was configured during/post installation), and the user will be notified. This may have a considerable performance impact on the operating system, though the degree of impact is dependent on how well the scanner was programmed. The goal is to stop any operations the malware may attempt on the system before they occur, including activities which might exploit bugs or trigger unexpected operating system behavior.

Anti-malware programs can combat malware in two ways:

1. They can provide real time protection against the installation of malware software on a computer. This type of malware protection works the same way as that of antivirus protection in that the anti-malware software scans all incoming network data for malware and blocks any threats it comes across.

2. Anti-malware software programs can be used solely for detection and removal of malware software that has already been installed onto a computer. This type of anti-malware software scans the contents of the Windows registry, operating system files, and installed programs on a computer and will provide a list of any threats found, allowing the user to choose which files to delete or keep, or to compare this list to a list of known malware components, removing files that match.

Real-time protection from malware works identically to real-time antivirus protection: the software scans disk files at download time, and blocks the activity of components known to represent malware. In some cases, it may also intercept attempts to install start-up items or to modify browser settings. Because many malware components are installed as a result of browser exploits or user error, using security software (some of which are anti-malware, though many are not) to "sandbox" browsers (essentially isolate the browser from the computer and hence any malware induced change) can also be effective in helping to restrict any damage done.

Examples of Microsoft Windows antivirus and anti-malware software include the optional Microsoft Security Essentials (for Windows XP, Vista, and Windows 7) for real-time protection, the Windows Malicious Software Removal Tool (now included with Windows (Security) Updates on "Patch Tuesday", the second Tuesday of each month), and Windows Defender (an optional download in the case of Windows XP, incorporating MSE functionality in the case of Windows 8 and later). Additionally, several capable antivirus software programs are available for free download from the Internet (usually restricted to non-commercial use). Tests found some free programs to be competitive with commercial ones. Microsoft's System File Checker can be used to check for and repair corrupted system files.

Some viruses disable System Restore and other important Windows tools such as Task Manager and Command Prompt. Many such viruses can be removed by rebooting the computer, entering Windows safe mode with networking, and then using system tools or Microsoft Safety Scanner.

Hardware implants can be of any type, so there can be no general way to detect them.

### Website Security Scans

As malware also harms the compromised websites (by breaking reputation, blacklisting in search engines, etc.), some websites offer vulnerability scanning. Such scans check the website, detect malware, may note outdated software, and may report known security issues.

### "Air Gap" Isolation or "Parallel Network"

As a last resort, computers can be protected from malware, and infected computers can be prevented from disseminating trusted information, by imposing an "air gap" (i.e. completely disconnecting them from all other networks). However, malware can still cross the air gap in some situations. For example, removable media can carry malware across the gap. In December 2013 researchers in Germany showed one way that an apparent air gap can be defeated.

"AirHopper", "BitWhisper", "GSMem" and "Fansmitter" are four techniques introduced by researchers that can leak data from air-gapped computers using electromagnetic, thermal and acoustic emissions.

### Grayware

Grayware is a term applied to unwanted applications or files that are not classified as malware, but can worsen the performance of computers and may cause security risks.

It describes applications that behave in an annoying or undesirable manner, and yet are less serious or troublesome than malware. Grayware encompasses spyware, adware, fraudulent dialers, joke programs, remote access tools and other unwanted programs that harm the performance of computers or cause inconvenience. The term came into use around 2004.

Another term, PUP, which stands for *Potentially Unwanted Program* (or PUA *Potentially Unwanted Application*), refers to applications that would be considered unwanted despite often having been downloaded by the user, possibly after failing to read a download agreement. PUPs include spyware, adware, and fraudulent dialers. Many security products classify unauthorised key generators as grayware, although they frequently carry true malware in addition to their ostensible purpose.

Software maker Malwarebytes lists several criteria for classifying a program as a PUP. Some adware (using stolen certificates) disables anti-malware and virus protection; technical remedies are available.

## History of Viruses and Worms

Before Internet access became widespread, viruses spread on personal computers by infecting the executable boot sectors of floppy disks. By inserting a copy of itself into the machine code instructions in these executables, a virus causes itself to be run whenever a program is run or the disk is booted. Early computer viruses were written for the Apple II and Macintosh, but they became more widespread with the dominance of the IBM PC and MS-DOS system. Executable-infecting viruses are dependent on users exchanging software or boot-able floppies and thumb drives so they spread rapidly in computer hobbyist circles.

The first worms, network-borne infectious programs, originated not on personal computers, but on multitasking Unix systems. The first well-known worm was the Internet Worm of 1988, which infected SunOS and VAX BSD systems. Unlike a virus, this worm did not insert itself into other programs. Instead, it exploited security holes (vulnerabilities) in network server programs and started itself running as a separate process. This same behavior is used by today's worms as well.

With the rise of the Microsoft Windows platform in the 1990s, and the flexible macros of its applications, it became possible to write infectious code in the macro language of Microsoft Word and similar programs. These *macro viruses* infect documents and templates rather than applications (executables), but rely on the fact that macros in a Word document are a form of executable code.

Today, worms are most commonly written for the Windows OS, although a few like Mare-D and the L1on worm are also written for Linux and Unix systems. Worms today work in the same basic way as 1988's Internet Worm: they scan the network and use vulnerable computers to replicate. Because they need no human intervention, worms can spread with incredible speed. The SQL Slammer infected thousands of computers in a few minutes in 2003.

## Academic Research

The notion of a self-reproducing computer program can be traced back to initial theories about the operation of complex automata. John von Neumann showed that in theory a program could reproduce itself. This constituted a plausibility result in computability theory. Fred Cohen experimented with computer viruses and confirmed Neumann's postulate and investigated other properties of malware such as detectability and self-obfuscation using rudimentary encryption. His doctoral dissertation was on the subject of computer viruses. The combination of cryptographic technology as part of the payload of the virus, exploiting it for attack purposes was initialized and investigated from the mid 1990s, and includes initial ransomware and evasion ideas.

# Mobile Malware

Mobile malware is malicious software that targets mobile phones or wireless-enabled Personal digital assistants (PDA), by causing the collapse of the system and loss or leakage of confidential information. As wireless phones and PDA networks have become more and more common and have grown in complexity, it has become increasingly difficult to ensure their safety and security against electronic attacks in the form of viruses or other malware.

## History

Cell phone malware were initially demonstrated by Brazilian software engineer Marcos Velasco. He created a virus that could be used by anyone in order to educate the public of the threat.

The first known mobile virus, "Timofonica", originated in Spain and was identified by antivirus labs in Russia and Finland in June 2000. "Timofonica" sent SMS messages to GSM mobile phones that read (in Spanish) "Information for you: Telefónica is fooling you." These messages were sent through the Internet SMS gate of the MoviStar mobile operator.

In June 2004, it was discovered that a company called Ojam had engineered an anti-piracy Trojan virus in older versions of its mobile phone game, *Mosquito*. This virus sent SMS text messages to the company without the user's knowledge. Although this malware was removed from the game's more recent versions, it still exists in older, unlicensed versions, and these may still be distributed on file-sharing networks and free software download web sites.

In July 2004, computer hobbyists released a proof-of-concept mobile virus *Cabir,* that replicates and spreads itself on Bluetooth wireless networks and infects mobile phones running the Symbian OS.

In March 2005, it was reported that a computer worm called Commwarrior-A had been infecting Symbian series 60 mobile phones. This specific worm replicated itself through the phone's Multimedia Messaging Service (MMS), sending copies of itself to other phone owners listed in the phone user's address book. Although the worm is not considered harmful, experts agree that it heralded a new age of electronic attacks on mobile phones.

In August 2010, Kaspersky Lab reported a trojan designated Trojan-SMS.AndroidOS.FakePlayer.a. This was the first malicious program classified as a Trojan SMS that affects smartphones running on Google's Android operating system, and which had already infected a number of mobile devices, sending SMS messages to premium rate numbers without the owner's knowledge or consent, and accumulating huge bills.

Currently, various antivirus software companies like Trend Micro, AVG, avast!, Comodo, Kaspersky Lab, PSafe, and Softwin are working to adapt their programs to the mobile operating systems that are most at risk. Meanwhile, operating system developers try to curb the spread of infections with quality control checks on software and content offered through their digital application distribution platforms, such as Google Play or Apple's App Store. Recent studies however show that mobile antivirus programs are ineffective due to the rapid evolution of mobile malware.

### Taxonomy

Four types of the most common malicious programs are known to affect mobile devices:

- Expander: Expanders target mobile meters for additional phone billing and profit.
- Worm: The main objective of this stand-alone type of malware is to endlessly reproduce itself and spread to other devices. Worms may also contain harmful and misleading instructions. Mobile worms may be transmitted via text messages SMS or MMS and typically do not require user interaction for execution.
- Trojan: Unlike worms, a Trojan horse always requires user interaction to be activated. This kind of virus is usually inserted into seemingly attractive and non-malicious executable files or applications that are downloaded to the device and executed by the user. Once activated, the malware can cause serious damage by infecting and deactivating other applications or the phone itself, rendering it paralyzed after a certain period of time or a certain number of operations. Usurpation data (spyware) synchronizes with calendars, email accounts, notes, and any other source of information before it is sent to a remote server.
- Spyware: This malware poses a threat to mobile devices by collecting, using, and spreading a user's personal or sensitive information without the user's consent or knowledge. It is mostly classified into four categories: system monitors, trojans, adware, and tracking cookies.
- Ghost Push: This is a kind of malware which infects the Android OS by automatically gaining root access, downloading malicious software, converting to a system app and then losing root access which makes it virtually impossible to remove the infection with a factory reset unless the firmware is reflashed. The malware hogs all system resources making it unresponsive and drains the battery. The advertisements appear anytime either in full screen, as part of a display, or in the status bar. The unnecessary apps automatically activate and sometimes download more malicious software when connected to the internet. It is hard to detect and remove. It steals the personal data of the user from the phone.

### Notable Mobile Malicious Programs

- Cabir: This malware infects mobile phones running on Symbian OS and was first identified in June 2004. When a phone is infected, the message 'Caribe' is displayed on the phone's screen and is displayed every time the phone is turned on. The worm then attempts to spread to other phones in the area using wireless Bluetooth signals, although the recipient has to confirm this manually.
- Duts: This parasitic file infector virus is the first known virus for the Pocket PC platform. It attempts to infect all EXE files that are larger than 4096 bytes in the current directory.

- Skulls: A trojan horse piece of code that targets mainly Symbian OS. Once downloaded, the virus replaces all phone desktop icons with images of a skull. It also renders all phone applications useless. This malware also tends to mass text messages containing malicious links to all contacts accessible through the device in order to spread the damage. This mass texting can also give rise to high expenses.
- Commwarrior: This malware was identified in 2005. It was the first worm to use MMS messages and can spread through Bluetooth as well. It infects devices running under OS Symbian Series 60. The executable worm file, once launched, hunts for accessible Bluetooth devices and sends the infected files under a random name to various devices.
- Gingermaster: A trojan developed for an Android platform that propagates by installing applications that incorporate a hidden malware for installation in the background. It exploits the frailty in the version Gingerbread (2.3) of the operating system to use super-user permissions by privileged escalation. It then creates a service that steals information from infected terminals (user ID, number SIM, phone number, IMEI, IMSI, screen resolution and local time) by sending it to a remote server through petitions HTTP.
- DroidKungFu: A trojan content in Android applications, which when executed, obtains root privileges and installs the file com.google. ssearch.apk, which contains a back door that allows files to be removed, open home pages to be supplied, and 'open web and download and install' application packages. This virus collects and sends to a remote server all available data on the terminal.
- Ikee: The first worm known for iOS platforms. It only works on terminals that were previously made a process of jailbreak, and spreads by trying to access other devices using the SSH protocol, first through the subnet that is connected to the device. Then, it repeats the process generating a random range and finally uses some preset ranges corresponding to the IP address of certain telephone companies. Once the computer is infected, the wallpaper is replaced by a photograph of the singer Rick Astley, a reference to the Rickroll phenomenon.
- Gunpoder : This worm file infector virus is the first known virus that officially infected the Google Play Store in few countries, including Brazil.
- Shedun: adware serving malware able to root Android devices.
- HummingBad - has infected over 10 million Android operating systems. User details are sold and adverts are tapped on without the user's knowledge thereby generating fraudulent advertising revenue.

## Dendroid

Dendroid is malware that affects Android OS and targets the mobile platform.

It was first discovered in early of 2014 by Symantec and appeared in the underground for sale mi-for $300. Some things were noted in Dendroid, such as being able to hide from emulators at the time. When first discovered in 2014 it was one of the most sophisticated Android remote administration tools known at that time. It was one of the first Trojan applications to get past Google's Bouncer and caused researchers to warn about it being easier to create Android malware due to it. It also seems to have follow in the footsteps of Zeus and SpyEye by having simple-to-use command and control panels. The code appeared to be leaked somewhere around 2014. It was noted that an apk binder was included in the leak, which provided a simple way to bind Dendroid to legitimate applications.

It is capable of:

- Deleting call logs.
- Opening web pages.
- Dialing any number.
- Recording calls.
- SMS intercepting.
- Uploading images and video.
- Opening an application.
- Performing denial-of-service attacks.
- Changing the command and control server.

## Spyware

Spyware is software that aims to gather information about a person or organization without their knowledge and that may send such information to another entity without the consumer's consent, or that asserts control over a computer without the consumer's knowledge.

"Spyware" is mostly classified into four types: system monitors, trojans, adware, and tracking cookies. Spyware is mostly used for the purposes of tracking and storing Internet users' movements on the Web and serving up pop-up ads to Internet users.

Whenever spyware is used for malicious purposes, its presence is typically hidden from the user and can be difficult to detect. Some spyware, such as keyloggers, may be installed by the owner of a shared, corporate, or public computer intentionally in order to monitor users.

While the term *spyware* suggests software that monitors a user's computing, the functions of spyware can extend beyond simple monitoring. Spyware can collect almost any type of data, including personal information like internet surfing habits, user logins, and bank or credit account information. Spyware can also interfere with user control of a computer by installing additional software or redirecting web browsers. Some spyware can change computer settings, which can result in slow Internet connection speeds, un-authorized changes in browser settings, or changes to software settings.

Sometimes, spyware is included along with genuine software, and may come from a malicious website or may have been added to the intentional functionality of genuine software. In response to the emergence of spyware, a small industry has sprung up dealing in anti-spyware software. Running anti-spyware software has become a widely recognized element of computer security practices, especially for computers running Microsoft Windows. A number of jurisdictions have passed anti-spyware laws, which usually target any software that is surreptitiously installed to control a user's computer.

In German-speaking countries, spyware used or made by the government is called *govware* by computer experts (in common parlance: *Regierungstrojaner*, literally 'Government Trojan'). Govware is typically a trojan horse software used to intercept communications from the target computer. Some countries like Switzerland and Germany have a legal framework governing the use of such software. In the US, the term policeware has been used for similar purposes.

Use of the term "spyware" has eventually declined as the practice of tracking users has been pushed ever further into the mainstream by major websites and data mining companies; these generally break no known laws and compel users to be tracked, not by fraudulent practices per se, but by the default settings created for users and the language of terms-of-service agreements. As one documented example, on March 7, 2011, CBS/Cnet News reported on a Wall Street Journal analysis revealing the practice of Facebook and other websites of tracking users' browsing activity, linked to their identity, far beyond users' visit and activity within the Facebook site itself. The report stated "Here's how it works. You go to Facebook, you log in, you spend some time there, and then ... you move on without logging out. Let's say the next site you go to is New York Times. Those buttons, without you clicking on them, have just reported back to Facebook and Twitter that you went there and also your identity within those accounts. Let's say you moved on to something like a site about depression. This one also has a tweet button, a Google widget, and those, too, can report back who you are and that you went there." The WSJ analysis was researched by Brian Kennish, founder of Disconnect, Inc.

## Routes of Infection

Spyware does not necessarily spread in the same way as a virus or worm because infected systems generally do not attempt to transmit or copy the software to other computers. Instead, spyware installs itself on a system by deceiving the user or by exploiting software vulnerabilities.

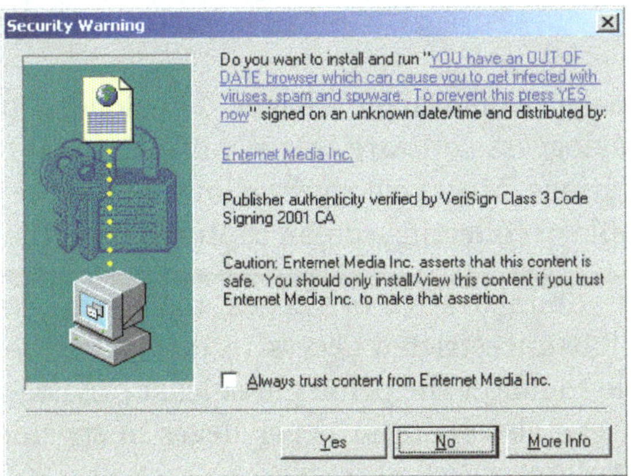


Malicious websites attempt to install spyware on readers' computers.

Most spyware is installed without knowledge, or by using deceptive tactics. Spyware may try to deceive users by bundling itself with desirable software. Other common tactics are using a Trojan horse, spy gadgets that look like normal devices but turn out to be something else, such as a USB Keylogger. These devices actually are connected to the device as memory units but are capable of recording each stroke made on the keyboard. Some spyware authors infect a system through security holes in the Web browser or in other software. When the user navigates to a Web page controlled by the spyware author, the page contains code which attacks the browser and forces the download and installation of spyware.

The installation of spyware frequently involves Internet Explorer. Its popularity and history of security issues have made it a frequent target. Its deep integration with the Windows environment make it susceptible to attack into the Windows operating system. Internet Explorer also serves as a point of attachment for spyware in the form of Browser Helper Objects, which modify the browser's behavior to add toolbars or to redirect traffic.

## Effects and Behaviors

A spyware program is rarely alone on a computer: an affected machine usually has multiple infections. Users frequently notice unwanted behavior and degradation of system performance. A spyware infestation can create significant unwanted CPU activity, disk usage, and network traffic. Stability issues, such as applications freezing, failure to boot, and system-wide crashes are also common. Spyware, which interferes with networking software, commonly causes difficulty connecting to the Internet.

In some infections, the spyware is not even evident. Users assume in those situations that the performance issues relate to faulty hardware, Windows installation problems, or another infection. Some owners of badly infected systems resort to contacting technical support experts, or even buying a new computer because the existing system "has become too slow". Badly infected systems may require a clean reinstallation of all their software in order to return to full functionality.

Moreover, some types of spyware disable software firewalls and anti-virus software, and/or reduce browser security settings, which further open the system to further opportunistic infections. Some spyware disables or even removes competing spyware programs, on the grounds that more spyware-related annoyances make it even more likely that users will take action to remove the programs.

Keyloggers are sometimes part of malware packages downloaded onto computers without the owners' knowledge. Some keyloggers software is freely available on the internet while others are commercial or private applications. Most keyloggers allow not only keyboard keystrokes to be captured but also are often capable of collecting screen captures from the computer.

A typical Windows user has administrative privileges, mostly for convenience. Because of this, any program the user runs has unrestricted access to the system. As with other operating systems, Windows users are able to follow the principle of least privilege and use non-administrator accounts. Alternatively, they can also reduce the privileges of specific vulnerable Internet-facing processes such as Internet Explorer.

Since Windows Vista, by default, a computer administrator runs everything under limited user privileges. When a program requires administrative privileges, a User Account Control pop-up will prompt the user to allow or deny the action. This improves on the design used by previous versions of Windows.

## Remedies and Prevention

As the spyware threat has worsened, a number of techniques have emerged to counteract it. These include programs designed to remove or block spyware, as well as various user practices which reduce the chance of getting spyware on a system.

Nonetheless, spyware remains a costly problem. When a large number of pieces of spyware have infected a Windows computer, the only remedy may involve backing up user data, and fully re-installing the operating system. For instance, some spyware cannot be completely removed by Symantec, Microsoft, PC Tools.

## Anti-spyware Programs

Many programmers and some commercial firms have released products dedicated to remove or block spyware. Programs such as PC Tools' Spyware Doctor, Lavasoft's *Ad-Aware SE* and Patrick Kolla's *Spybot - Search & Destroy* rapidly gained popularity as tools to remove, and in some cases intercept, spyware programs. On December 16, 2004, Microsoft acquired the *GIANT AntiSpyware* software, rebranding it as *Windows AntiSpyware beta* and releasing it as a free download for Genuine Windows XP and Windows 2003 users. (In 2006 it was renamed Windows Defender).

Major anti-virus firms such as Symantec, PC Tools, McAfee and Sophos have also added anti-spyware features to their existing anti-virus products. Early on, anti-virus firms expressed reluctance to add anti-spyware functions, citing lawsuits brought by spyware authors against the authors of web sites and programs which described their products as "spyware". However, recent versions of these major firms' home and business anti-virus products do include anti-spyware functions, albeit treated differently from viruses. Symantec Anti-Virus, for instance, categorizes spyware programs as "extended threats" and now offers real-time protection against these threats.

## How anti-spyware Software Works

Anti-spyware programs can combat spyware in two ways:

1. They can provide real-time protection in a manner similar to that of anti-virus protection: they scan all incoming network data for spyware and blocks any threats it detects.

2. Anti-spyware software programs can be used solely for detection and removal of spyware software that has already been installed into the computer. This kind of anti-spyware can often be set to scan on a regular schedule.

Such programs inspect the contents of the Windows registry, operating system files, and installed programs, and remove files and entries which match a list of known spyware. Real-time protection from spyware works identically to real-time anti-virus protection: the software scans disk files at download time, and blocks the activity of components known to represent spyware. In some cases, it may also intercept attempts to install start-up items or to modify browser settings. Earlier versions of anti-spyware programs focused chiefly on detection and removal. Javacool Software's SpywareBlaster, one of the first to offer real-time protection, blocked the installation of ActiveX-based spyware.

Like most anti-virus software, many anti-spyware/adware tools require a frequently updated database of threats. As new spyware programs are released, anti-spyware developers discover and evaluate them, adding to the list of known spyware, which allows the software to detect and remove new spyware. As a result, anti-spyware software is of limited usefulness without regular updates. Updates may be installed automatically or manually.

A popular generic spyware removal tool used by those that requires a certain degree of expertise is HijackThis, which scans certain areas of the Windows OS where spyware often resides and presents a list with items to delete manually. As most of the items are legitimate windows files/registry entries it is advised for those who are less knowledgeable on this subject to post a HijackThis log on the numerous antispyware sites and let the experts decide what to delete.

If a spyware program is not blocked and manages to get itself installed, it may resist attempts to terminate or uninstall it. Some programs work in pairs: when an anti-spyware scanner (or the user) terminates one running process, the other one respawns the killed program. Likewise, some spyware will detect attempts to remove registry keys and immediately add them again. Usually, booting the infected computer in safe mode allows an anti-spyware program a better chance of removing persistent spyware. Killing the process tree may also work.

## Security Practices

To detect spyware, computer users have found several practices useful in addition to installing anti-spyware programs. Many users have installed a web browser other than Internet Explorer, such as Mozilla Firefox or Google Chrome. Though no browser is completely safe, Internet Explorer was once at a greater risk for spyware infection due to its large user base as well as vulnerabilities such as ActiveX but these three major browsers are now close to equivalent when it comes to security.

Some ISPs—particularly colleges and universities—have taken a different approach to blocking spyware: they use their network firewalls and web proxies to block access to Web sites known to install spyware. On March 31, 2005, Cornell University's Information Technology department released a report detailing the behavior of one particular piece of proxy-based spyware, *Marketscore*, and the steps the university took to intercept it. Many other educational institutions have taken similar steps.

Individual users can also install firewalls from a variety of companies. These monitor the flow of information going to and from a networked computer and provide protection against spyware and malware. Some users install a large hosts file which prevents the user's computer from connecting to known spyware-related web addresses. Spyware may get installed via certain shareware programs offered for download. Downloading programs only from reputable sources can provide some protection from this source of attack.

## Applications

### "Stealware" and Affiliate Fraud

A few spyware vendors, notably 180 Solutions, have written what the *New York Times* has dubbed "stealware", and what spyware researcher Ben Edelman terms *affiliate fraud*, a form of click fraud. Stealware diverts the payment of affiliate marketing revenues from the legitimate affiliate to the spyware vendor.

Spyware which attacks affiliate networks places the spyware operator's affiliate tag on the user's activity – replacing any other tag, if there is one. The spyware operator is the only party that gains from this. The user has their choices thwarted, a legitimate affiliate loses revenue, networks' reputations are injured, and vendors are harmed by having to pay out affiliate revenues to an "affiliate" who is not party to a contract. Affiliate fraud is a violation of the terms of service of most affiliate marketing networks. As a result, spyware operators such as 180 Solutions have been terminated from affiliate networks including LinkShare and ShareSale. Mobile devices can also be vulnerable to chargeware, which manipulates users into illegitimate mobile charges.

### Identity Theft and Fraud

In one case, spyware has been closely associated with identity theft. In August 2005, researchers from security software firm Sunbelt Software suspected the creators of the common CoolWebSearch spyware had used it to transmit "chat sessions, user names, passwords, bank information, etc."; however it turned out that "it actually (was) its own sophisticated criminal little trojan that's independent of CWS." This case is currently under investigation by the FBI.

The Federal Trade Commission estimates that 27.3 million Americans have been victims of identity theft, and that financial losses from identity theft totaled nearly $48 billion for businesses and financial institutions and at least $5 billion in out-of-pocket expenses for individuals.

### Digital Rights Management

Some copy-protection technologies have borrowed from spyware. In 2005, Sony BMG Music Entertainment was found to be using rootkits in its XCP digital rights management technology Like spyware, not only was it difficult to detect and uninstall, it was so poorly written that most efforts to remove it could have rendered computers unable to function. Texas Attorney General Greg Abbott filed suit, and three separate class-action suits were filed. Sony BMG later provided a workaround on its website to help users remove it.

Beginning on April 25, 2006, Microsoft's Windows Genuine Advantage Notifications application was installed on most Windows PCs as a "critical security update". While the main purpose of this deliberately uninstallable application is to ensure the copy of Windows on the machine was lawfully purchased and installed, it also installs software that has been accused of "phoning home" on a daily basis, like spyware. It can be removed with the RemoveWGA tool.

### Personal Relationships

Spyware has been used to monitor electronic activities of partners in intimate relationships. At least one software package, Loverspy, was specifically marketed for this purpose. Depending on local laws regarding communal/marital property, observing a partner's online activity without their consent may be illegal; the author of Loverspy and several users of the product were indicted in California in 2005 on charges of wiretapping and various computer crimes.

## Browser Cookies

Anti-spyware programs often report Web advertisers' HTTP cookies, the small text files that track browsing activity, as spyware. While they are not always inherently malicious, many users object to third parties using space on their personal computers for their business purposes, and many anti-spyware programs offer to remove them.

## Examples

These common spyware programs illustrate the diversity of behaviors found in these attacks. Note that as with computer viruses, researchers give names to spyware programs which may not be used by their creators. Programs may be grouped into "families" based not on shared program code, but on common behaviors, or by "following the money" of apparent financial or business connections. For instance, a number of the spyware programs distributed by Claria are collectively known as "Gator". Likewise, programs that are frequently installed together may be described as parts of the same spyware package, even if they function separately.

- CoolWebSearch, a group of programs, takes advantage of Internet Explorer vulnerabilities. The package directs traffic to advertisements on Web sites including coolwebsearch.com. It displays pop-up ads, rewrites search engine results, and alters the infected computer's hosts file to direct DNS lookups to these sites.
- FinFisher, sometimes called FinSpy is a high-end surveillance suite sold to law enforcement and intelligence agencies. Support services such as training and technology updates are part of the package.
- HuntBar, aka WinTools or Adware.Websearch, was installed by an ActiveX drive-by download at affiliate Web sites, or by advertisements displayed by other spyware programs—an example of how spyware can install more spyware. These programs add toolbars to IE, track aggregate browsing behavior, redirect affiliate references, and display advertisements.
- Internet Optimizer, also known as DyFuCa, redirects Internet Explorer error pages to advertising. When users follow a broken link or enter an erroneous URL, they see a page of advertisements. However, because password-protected Web sites (HTTP Basic authentication) use the same mechanism as HTTP errors, Internet Optimizer makes it impossible for the user to access password-protected sites.
- Spyware such as Look2Me hides inside system-critical processes and start up even in safe mode. With no process to terminate they are harder to detect and remove, which is a combination of both spyware and a rootkit. Rootkit technology is also seeing increasing use, as newer spyware programs also have specific countermeasures against well known anti-malware products and may prevent them from running or being installed, or even uninstall them.
- Movieland, also known as Moviepass.tv and Popcorn.net, is a movie download service that

has been the subject of thousands of complaints to the Federal Trade Commission (FTC), the Washington State Attorney General's Office, the Better Business Bureau, and other agencies. Consumers complained they were held hostage by a cycle of oversized pop-up windows demanding payment of at least $29.95, claiming that they had signed up for a three-day free trial but had not cancelled before the trial period was over, and were thus obligated to pay. The FTC filed a complaint, since settled, against Movieland and eleven other defendants charging them with having "engaged in a nationwide scheme to use deception and coercion to extract payments from consumers."

- WeatherStudio has a plugin that displays a window-panel near the bottom of a browser window. The official website notes that it is easy to remove (uninstall) WeatherStudio from a computer, using its own uninstall-program, such as under C:\Program Files\WeatherStudio. Once WeatherStudio is removed, a browser returns to the prior display appearance, without the need to modify the browser settings.
- Zango (formerly 180 Solutions) transmits detailed information to advertisers about the Web sites which users visit. It also alters HTTP requests for affiliate advertisements linked from a Web site, so that the advertisements make unearned profit for the 180 Solutions company. It opens pop-up ads that cover over the Web sites of competing companies (as seen in their [Zango End User License Agreement]).
- Zlob trojan, or just Zlob, downloads itself to a computer via an ActiveX codec and reports information back to Control Server. Some information can be the search-history, the Websites visited, and even keystrokes. More recently, Zlob has been known to hijack routers set to defaults.

## History and Development

The first recorded use of the term spyware occurred on October 16, 1995 in a Usenet post that poked fun at Microsoft's business model. *Spyware* at first denoted *software* meant for espionage purposes. However, in early 2000 the founder of Zone Labs, Gregor Freund, used the term in a press release for the ZoneAlarm Personal Firewall. Later in 2000, a parent using ZoneAlarm was alerted to the fact that "Reader Rabbit," educational software marketed to children by the Mattel toy company, was surreptitiously sending data back to Mattel. Since then, "spyware" has taken on its present sense.

According to a 2005 study by AOL and the National Cyber-Security Alliance, 61 percent of surveyed users' computers were infected with form of spyware. 92 percent of surveyed users with spyware reported that they did not know of its presence, and 91 percent reported that they had not given permission for the installation of the spyware. As of 2006, spyware has become one of the preeminent security threats to computer systems running Microsoft Windows operating systems. Computers on which Internet Explorer (IE) is the primary browser are particularly vulnerable to such attacks, not only because IE is the most widely used, but because its tight integration with Windows allows spyware access to crucial parts of the operating system.

Before Internet Explorer 6 SP2 was released as part of Windows XP Service Pack 2, the browser would automatically display an installation window for any ActiveX component that a website wanted to install. The combination of user ignorance about these changes, and the assumption by Internet Explorer that all ActiveX components are benign, helped to spread spyware significantly.

Many spyware components would also make use of exploits in JavaScript, Internet Explorer and Windows to install without user knowledge or permission.

The Windows Registry contains multiple sections where modification of key values allows software to be executed automatically when the operating system boots. Spyware can exploit this design to circumvent attempts at removal. The spyware typically will link itself from each location in the registry that allows execution. Once running, the spyware will periodically check if any of these links are removed. If so, they will be automatically restored. This ensures that the spyware will execute when the operating system is booted, even if some (or most) of the registry links are removed.

## Programs Distributed with Spyware

- Kazaa,
- Morpheus,
- WeatherBug,
- WildTangent.

## Programs Formerly Distributed with Spyware

- AOL Instant Messenger (AOL Instant Messenger still packages Viewpoint Media Player, and WildTangent),
- DivX,
- FlashGet,
- magicJack.

## Rogue Anti-spyware Programs

Malicious programmers have released a large number of rogue (fake) anti-spyware programs, and widely distributed Web banner ads can warn users that their computers have been infected with spyware, directing them to purchase programs which do not actually remove spyware—or else, may add more spyware of their own.

The recent proliferation of fake or spoofed antivirus products that bill themselves as antispyware can be troublesome. Users may receive popups prompting them to install them to protect their computer, when it will in fact add spyware. This software is called rogue software. It is recommended that users do not install any freeware claiming to be anti-spyware unless it is verified to be legitimate. Some known offenders include:

- AntiVirus 360,
- Antivirus 2009,
- AntiVirus Gold,
- ContraVirus,
- MacSweeper,

- Pest Trap,
- PSGuard,
- Spy Wiper,
- Spydawn,
- Spylocked,
- Spysheriff,
- SpyShredder,
- Spyware Quake,
- SpywareStrike,
- UltimateCleaner,
- WinAntiVirus Pro 2006,
- Windows Police Pro,
- WinFixer,
- WorldAntiSpy.

Fake antivirus products constitute 15 percent of all malware.

On January 26, 2006, Microsoft and the Washington state attorney general filed suit against Secure Computer for its Spyware Cleaner product.

## Legal Issues

### Criminal Law

Unauthorized access to a computer is illegal under computer crime laws, such as the U.S. Computer Fraud and Abuse Act, the U.K.'s Computer Misuse Act, and similar laws in other countries. Since owners of computers infected with spyware generally claim that they never authorized the installation, a *prima facie* reading would suggest that the promulgation of spyware would count as a criminal act. Law enforcement has often pursued the authors of other malware, particularly viruses. However, few spyware developers have been prosecuted, and many operate openly as strictly legitimate businesses, though some have faced lawsuits.

Spyware producers argue that, contrary to the users' claims, users do in fact give consent to installations. Spyware that comes bundled with shareware applications may be described in the legalese text of an end-user license agreement (EULA). Many users habitually ignore these purported contracts, but spyware companies such as Claria say these demonstrate that users have consented.

Despite the ubiquity of EULAs agreements, under which a single click can be taken as consent to the entire text, relatively little caselaw has resulted from their use. It has been established in most common law jurisdictions that this type of agreement can be a binding contract *in certain circumstances*. This does not, however, mean that every such agreement is a contract, or that every term in one is enforceable.

Some jurisdictions, including the U.S. states of Iowa and Washington, have passed laws criminalizing some forms of spyware. Such laws make it illegal for anyone other than the owner or operator of a computer to install software that alters Web-browser settings, monitors keystrokes, or disables computer-security software.

In the United States, lawmakers introduced a bill in 2005 entitled the Internet Spyware Prevention Act, which would imprison creators of spyware.

## Administrative Sanctions

### US FTC Actions

The US Federal Trade Commission has sued Internet marketing organizations under the "unfairness doctrine" to make them stop infecting consumers' PCs with spyware. In one case, that against Seismic Entertainment Productions, the FTC accused the defendants of developing a program that seized control of PCs nationwide, infected them with spyware and other malicious software, bombarded them with a barrage of pop-up advertising for Seismic's clients, exposed the PCs to security risks, and caused them to malfunction. Seismic then offered to sell the victims an "antispyware" program to fix the computers, and stop the popups and other problems that Seismic had caused. On November 21, 2006, a settlement was entered in federal court under which a $1.75 million judgment was imposed in one case and $1.86 million in another, but the defendants were insolvent.

In a second case, brought against CyberSpy Software LLC, the FTC charged that CyberSpy marketed and sold "RemoteSpy" keylogger spyware to clients who would then secretly monitor unsuspecting consumers' computers. According to the FTC, Cyberspy touted RemoteSpy as a "100% undetectable" way to "Spy on Anyone. From Anywhere." The FTC has obtained a temporary order prohibiting the defendants from selling the software and disconnecting from the Internet any of their servers that collect, store, or provide access to information that this software has gathered. The case is still in its preliminary stages. A complaint filed by the Electronic Privacy Information Center (EPIC) brought the RemoteSpy software to the FTC's attention.

### Netherlands OPTA

An administrative fine, the first of its kind in Europe, has been issued by the Independent Authority of Posts and Telecommunications (OPTA) from the Netherlands. It applied fines in total value of Euro 1,000,000 for infecting 22 million computers. The spyware concerned is called DollarRevenue. The law articles that have been violated are art. 4.1 of the Decision on universal service providers and on the interests of end users; the fines have been issued based on art. 15.4 taken together with art. 15.10 of the Dutch telecommunications law.

## Civil Law

Former New York State Attorney General and former Governor of New York Eliot Spitzer has pursued spyware companies for fraudulent installation of software. In a suit brought in 2005 by Spitzer, the California firm Intermix Media, Inc. ended up settling, by agreeing to pay US$7.5 million and to stop distributing spyware.

The hijacking of Web advertisements has also led to litigation. In June 2002, a number of large Web publishers sued Claria for replacing advertisements, but settled out of court.

Courts have not yet had to decide whether advertisers can be held liable for spyware that displays their ads. In many cases, the companies whose advertisements appear in spyware pop-ups do not directly do business with the spyware firm. Rather, they have contracted with an advertising agency, which in turn contracts with an online subcontractor who gets paid by the number of "impressions" or appearances of the advertisement. Some major firms such as Dell Computer and Mercedes-Benz have sacked advertising agencies that have run their ads in spyware.

### Libel Suits by Spyware Developers

Litigation has gone both ways. Since "spyware" has become a common pejorative, some makers have filed libel and defamation actions when their products have been so described. In 2003, Gator (now known as Claria) filed suit against the website PC Pitstop for describing its program as "spyware". PC Pitstop settled, agreeing not to use the word "spyware", but continues to describe harm caused by the Gator/Claria software. As a result, other anti-spyware and anti-virus companies have also used other terms such as "potentially unwanted programs" or greyware to denote these products.

### WebcamGate

In the 2010 WebcamGate case, plaintiffs charged two suburban Philadelphia high schools secretly spied on students by surreptitiously and remotely activating webcams embedded in school-issued laptops the students were using at home, and therefore infringed on their privacy rights. The school loaded each student's computer with LANrev's remote activation tracking software. This included the now-discontinued "TheftTrack". While TheftTrack was not enabled by default on the software, the program allowed the school district to elect to activate it, and to choose which of the TheftTrack surveillance options the school wanted to enable.

TheftTrack allowed school district employees to secretly remotely activate the webcam embedded in the student's laptop, above the laptop's screen. That allowed school officials to secretly take photos through the webcam, of whatever was in front of it and in its line of sight, and send the photos to the school's server. The LANrev software disabled the webcams for all other uses (*e.g.*, students were unable to use Photo Booth or video chat), so most students mistakenly believed their webcams did not work at all. In addition to webcam surveillance, TheftTrack allowed school officials to take screenshots, and send them to the school's server. In addition, LANrev allowed school officials to take snapshots of instant messages, web browsing, music playlists, and written compositions. The schools admitted to secretly snapping over 66,000 webshots and screenshots, including webcam shots of students in their bedrooms.

## Dirtbox

A dirtbox (or DRT box) is a cell site simulator, or a phone device mimicking a cell phone tower. The device is designed to create a signal strong enough within a short range that it forces dormant mobile phones to automatically switch over to it. It is used by the United States Marshals Service, mounted on aircraft all over the U.S., to detect and locate cell phones and thus collect information, and can be used to jam phones. The name stems from the company that originally developed it, Digital Receiver Technology, Inc., abbreviated DRT, owned by the Boeing company. Boeing describes the device as a hybrid of "jamming, managed access and detection". A similar device with a more limited range that has been widely used by United States federal entities, including the Federal Bureau of Investigation (FBI), is the controversial StingRay phone tracker.

## History

It is not known when Digital Receiver Technology, Inc. (DRT) first manufactured the dirtbox. DRT does not publicly advertise the dirtbox, stating, "Due to the sensitive nature of our work, we are unable to publicly advertise many of our products" on its Web site. *The Wall Street Journal* wrote that the program by the U.S. Marshals Service "fully matured by 2007". Boeing bought DRT in 2008.

Similar devices from the Harris Corporation, like the Stingray phone tracker, have been sold around the same time. Since 2008, their airborne mounting kit for cell phone surveillance has been said to cost $9,000.

On June 11, 2010, the Boeing Company had asked the National Telecommunications and Information Administration to advise the United States Congress that the "... Communications Act of 1934 be modified to allow prison officials and state and local law enforcement to use appropriate cell phone management". It suggested that special weapons and tactics (SWAT) teams and other paramilitary tactical units could control wireless communications in a building during a raid.

The Chicago Police Department bought dirt boxes to eavesdrop on demonstrators during the 2012 NATO summit and used them in 2014 Black Lives Matter demonstrations, and in 2015 it became known that the Los Angeles Police Department had purchased them too.

## Technology

The device was described as being 2-square-foot (0.19 $m^2$) in size in the first mass media publication that shed light on it.

It acts as a fake cell phone tower, and utilizes IMSI-catcher technology, which stands for "international mobile subscriber identity". Phone networks use this system to identify an individual subscriber. The device emits a pilot signal made to appear stronger than that from the service provider cell tower, which forces all phones within its range to broadcast their IMSI number and electronic serial number (ESN). Encryption does not prevent this process. Furthermore, the devices can retrieve the encryption session keys for a cellphone in less than a second with success rates of 50 to 75% (in real world conditions)".

By mounting the device on a plane, it can locate a phone within 10 feet, while another source claims that by triangulating flights, a dirtbox can identify a phone accurately up to two feet.

The dirtbox is a hybrid of detection, managed access and jamming technologies: According to "people with knowledge of the program", the device can determine which phones belong to suspects and which belong to non-suspects, and that "cell phones not of interest, such as those belonging to prison personnel or commercial users in the area, are returned to their local network."

The dirtbox can also selectively jam for brief intervals, that is, interrupt or disable calls on certain phones, preventing contraband use in prison. It can also retrieve data from a phone. The technology is "unobtrusive to legitimate wireless communications", as described by Boeing, and bypasses phone companies in its operations.

## Agency use

### Law Enforcement

As of November 2014, the U.S. Marshals Service Technical Operations Group has been using the device fixed on manned airplanes to track fugitives, but it can deploy it on "targets requested by other parts of the Justice Program". It operates from at least five U.S. airports "covering most of the U.S. population". It is unclear whether the U.S. Marshals Service is requesting court orders.

Specifically, the Marshals Service has used dirtboxes in the Mexican Drug War for tracking fugitives in coordination with Mexico's Naval Infantry Force and flights in Guatemala.

DRT boxes are used by the United States Special Operations Command, the Drug Enforcement Administration, the FBI and U.S. Customs and Border Protection. The U.S. Navy bought dirtboxes to mount on drones at Naval Air Weapons Station China Lake, its research and development facility in Southern California according to procurement documents. The Pentagon Washington Headquarters Services bought them in 2011.

### Signal Intelligence

Based on references to DRTBox in NSA's Boundless informant screenshots leaked by Edward Snowden, dirtbox is probably used by the NSA. Le Monde wrote in 2013, that "thanks to DRTBOX, 62.5 million phone data were collected in France". The United States Naval Special Warfare Development Group Group One bought a Digital Receiver Technology 1301B System on April 2, 2007 costing over $25,000 per the United States government procurement web site.

## U.S. Regulation

The National Telecommunications and Information Administration (NTIA) has known of dirtbox since at least 2010. In 2014, the United States Department of Justice refused to confirm or deny that the program exists, but an official said "it would be "utterly false" to conflate the law-enforcement program with the collection of bulk telephone records by the National Security Agency". The Federal Communications Commission, responsible for licensing and regulating cell-service providers, was not aware of this activity prior to The Wall Street Journal's exposé.

In January 2015 the US Senate Judiciary Committee asked Department of Justice and Department of Homeland Security which law enforcement agencies use the DRTbox,and the legal process/what policies existed to protect the privacy interests of those whose information is collected.

## Criticism

Dirtbox use has been criticized as violating the Fourth Amendment to the United States Constitution. including by U.S. Rep. Alan Grayson, D-Florida.

Brian Owsley, a law professor at Indiana Institute of Technology and former United States magistrate said "I think the government would need to obtain a search warrant based on probable cause consistent with the Fourth Amendment".

*The Guardian* quoted Michael German at New York University Law School, a former FBI agent, as saying: "The overriding problem is the excessive secrecy that hides the government's ever-expanding surveillance programs from public accountability."

Senator Edward Markey (Democrat, Massachusetts) and Al Franken (Democrat, Minnesota) have warned that Americans' privacy rights must be assured.

# Juice Jacking

Juice jacking is a term used to describe a cyber attack wherein malware might be installed on to, or data surreptitiously copied from, a smart phone, tablet or other computer device using a charging port that doubles as a data connection, typically over USB.

## Published Research

The Wall of Sheep, an event at Def_Con has set up and allowed public access to an informational juice jacking kiosk each year at DefCon since 2011. Their intent is to bring awareness of this attack to the general public. Each of the informational juice jacking kiosks set up at the Wall of Sheep village have included a hidden CPU which is used in some way to notify the user that they should not plug their devices in to public charging kiosks. The first informational juice jacking kiosk included a screen which would change from "Free charging station" to a warning message that the user "should not trust public charging stations with their devices". One of the researchers who designed the charging station for the Wall of Sheep has given public presentations which showcase more malicious acts which could be taken via the kiosk, such as data theft, device tracking and information on compromising existing charging kiosks.

Security researcher Kyle Osborn released an attack framework called P2P-ADB in 2012 which utilized USB On-The-Go to connect an attacker's phone to a target victim's device. This framework included examples and proof of concepts which would allow attackers to unlock locked phones, steal data from a phone including authentication keys granting the attacker access to the target device owner's google account.

Security researcher graduates and students from the Institute of Technology Georgia released a proof of concept malicious tool "Mactans" which utilized the USB charging port on Apple mobile devices at the 2013 Blackhat USA security briefings. They utilized inexpensive hardware components to construct a small sized malicious wall charger which could infect an iPhone with the then-current version of iOS with malicious software while it was being charged. The software could defeat any security measures built into iOS and mask itself in the same way Apple masks background processes in iOS.

Security researchers Karsten Nohl and Jakob Lell from srlabs published their research on BadUSB during the 2014 Blackhat USA security briefings. Their presentation on this attack mentions that a cellphone or tablet device charging on an infected computer would be one of the simplest method of propagating the BadUSB vulnerability. They include example malicious firmware code that would infect Android devices with BadUSB.

Researchers at Aries Security and the Wall of Sheep later revisited the juice jacking concept in 2016. They set up a "Video Jacking" charging station which was able to record the mirrored screen from phones plugged in to their malicious charging station. Affected devices at the time included android devices supporting SlimPort or MHL protocols over USB, as well as the most recent iPhone using a lightning charge cable connector.

## History

Brian Krebs was the first to report on this attack and coined the term "juice jacking". After seeing the informational cell phone charging kiosk set up in the Wall of Sheep at DefCon 19 in August 2011, he wrote the first article on his security journalism site Krebs on Security. The wall of sheep researchers including Brian Markus, Joseph Mlodzianowski and Robert Rowley designed the kiosk as an information tool to bring awareness to the potential attack vector and have discussed but not released tools publicly which perform malicious actions on the charging devices.

An episode of the hacking series Hak5 released in September 2012 showcased a number of attacks which can be utilized using an attack framework named P2P-ADB released by Kyle Osborn. The P2P-ADB attack framework discussed utilized one phone to attack another phone over a USB on the Go connection.

In late 2012 a document was released by the NSA warning government employees who travel about the threat of juice jacking and reminding the reader that during overseas travel only to use their personal power charging cables and not to charge in public kiosks or by utilizing other people's computers.

The Android Hackers Handbook released in March 2014 has dedicated sections discussing both juice jacking and the ADB-P2P framework.

Juice jacking was the central focus on an episode of *CSI: Cyber*. Season 1: Episode 9, "LoM1S" aired in April 2015.

## Mitigation

Apple's IOS has taken multiple security measures to reduce the attack surface over USB including no longer allowing the device to automatically mount as a hard drive when plugged in over USB, as well as release security patches for vulnerabilities such as those exploited by Mactans.

Android devices commonly prompt the user before allowing the device to be mounted as a hard drive when plugged in over USB. Since release 4.2.2, Android has implemented a whitelist verification step to disallow attackers to access the Android Debug Bridge without authorization.

Juice jacking is not possible if a device is charged via the AC adapter shipped with the device, a battery backup device, or by utilizing a USB cable that has its data cables removed. A tool originally called the USB Condom, and now renamed to SyncStop has been release with the sole purpose of disallowing data connections to be passed over a USB cable.

## Trusted Execution Environment

The Trusted Execution Environment (TEE) is a secure area of the main processor of a smart phone (or any connected device including tablets, set-top boxes and televisions). It guarantees code and data loaded inside to be protected with respect to confidentiality and integrity. The TEE as an isolated execution environment is providing security features such as isolated execution, integrity of Trusted Applications along with confidentiality of their assets. In general terms, the TEE offers an execution space that provides a higher level of security than a rich mobile operating system (mobile OS) and more functionality than a 'secure element' (SE).

Industry associations like GlobalPlatform (working to standardize specifications for the TEE) and Trusted Computing Group (working to align GlobalPlatform TEE specification with its Trusted Platform Module (TPM) technology for enhanced mobile security) have undertaken work in recent years.

### History

OMTP first defined the TEE in their 'Advanced Trusted Environment:OMTP TR1' standard, defining it as a "set of hardware and software components providing facilities necessary to support Applications" which had to meet the requirements of one of two defined security levels. The first security level, Profile 1, was targeted against only software attacks and whilst Profile 2, was targeted against both software and hardware attacks.

Commercial TEE solutions based on ARM TrustZone technology which conformed to the TR1 standard such as Trusted Foundations, developed by Trusted Logic, were later launched. This software would become part of the Trustonic joint venture, and the basis of future GlobalPlatform TEE solutions.

Work on the OMTP standards ended in mid 2010 when the group transitioned into the 'Wholesale Applications Community' (WAC).

The OMTP standards, including those defining a TEE, are hosted by GSMA.

In July 2010 GlobalPlatform first announced their own standardisation of the TEE, focusing first on the client API (the interface to the TEE within the mobile OS) which was expanded later to include the TEE internal API, and a compliance programme and standardised security level.

### Details

The TEE is an isolated environment that runs in parallel with the mobile OS, providing security for the rich environment. It is more secure than the mobile OS and offers a higher level of functionality than the SE, using a hybrid approach that utilizes both hardware and software to protect data. It therefore offers a level of security sufficient for most applications. Trusted applications running in a TEE have access to the full power of a device's main processor and memory, while hardware isolation protects these from user installed apps running in a main operating system. Software and cryptographic isolation inside the TEE protect the trusted applications contained within from each other.

Service providers, mobile network operators (MNO), operating system (OS) and application developers, device manufacturers, platform providers and silicon vendors are all key stakeholders are all interested in, and contributing to, the standardization efforts and will benefit from the resulting specifications.

## Uses

There are a number of use cases for the TEE:

### Premium Content Protection

The TEE is an ideal environment for protecting premium content (for example, HD films) on connected devices such as smart phones and HD televisions. Premium content is defined by its perceived value which is in itself defined by the quality of the material (4K high definition films are one example), the file's proximity to its release date (as content has more value the closer it is to its release) and by consumer recognition. The TEE is used to protect the highest value content and so will be deployed into devices where this content is available:

- 4k HD televisions.
- 4k HD set-top boxes.
- Tablet computers.
- Smartphones.

The TEE is used to protect the content once it is on the device. The content is encrypted during transmission or streaming so it is protected. The TEE protects the content once it has been decrypted on the device as it is a secure environment.

### Mobile Financial Services

As m-Commerce (mobile wallets, peer-to-peer payments, contactless payments and using a mobile device as a point of sale (POS) terminal) develops, stronger and more standardized mobile security is needed. In collaboration with near field communication (NFC) and SEs, the TEE needs to be deployed to ensure the device is secure and that consumers can carry out any financial transaction in a safe and trusted environment.

Sensitive mobile use cases often need some form of interaction with the end user, meaning that sensitive information needs to be 'exposed' in the mobile OS to the user for validation - 'to guarantee What You See Is What You Sign'. The TEE offers a safe and trusted user interface to enable authentication on a mobile device.

### Authentication

The TEE is ideal for supporting natural ID (facial recognition, fingerprint sensor and voice authorization) as PINs and passwords can be easily hacked and stolen. The authentication process is split into three stages:

- Extracting an 'image' (scanning the fingerprint or capturing a voice sample, for example).
- A reference 'template' stored on the device for comparison with the extracted 'image'.

- A match engine to process the comparison between the 'image' and the 'template'.

The TEE is an ideal area within a mobile device to house the match engine and the associated processes required to authenticate the user. The increased security of this environment is able to protect the data and establish a buffer against the non-secure apps located in mobile OS. This additional security will help to satisfy the needs of service providers in addition to keeping the costs low for handset developers.

The FIDO Alliance is collaborating with GlobalPlatform to standardize the TEE for natural ID implementations.

## Enterprise and Government

The TEE can be used by governments and enterprises to enable the secure handling of confidential information on a mobile device. The TEE offers a level of protection against software attacks generated in the mobile OS and assists in the control of access rights. It achieves this by housing sensitive, 'trusted' applications that need to be isolated and protected from the mobile OS and any malicious malware that may be present. Through utilizing the functionality and security levels offered by the TEE, governments and enterprises can be assured that employees using their own devices are doing so in a secure and trusted manner.

### Implementations

The following embedded hardware technologies can be used to support TEE implementations:

- AMD:
    - Platform Security Processor (PSP).
- ARM:
    - TrustZone.
- Intel x86-64 instruction set:
    - SGX Software Guard Extensions.

Several TEE implementations are available from different TEE providers:

- Commercial implementations:
    - Kinibi, a commercial implementation from Trustonic that has been qualified by GlobalPlatform.
    - securiTEE, a commercial implementation from Solacia that has been qualified by GlobalPlatform.
- Open-source implementations:
    - OP-TEE, an open source implementation under BSD license, originally from STMicroelectronics, now owned and maintained by Linaro.

  - TLK, an open-source implementation from Nvidia under BSD license.
  - T6, and open-source implementation and research topic under GPL license.
  - Open TEE, an open source implementation and research project from the University of Helsinki and sponsored by Intel. Provided under an Apache license.
- Implementations with dual commercial/open-source licensing:
  - SierraTEE, an implementation from Sierraware available both under commercial and GPL-licensing.

## Standardization

While there are a number of proprietary systems, GlobalPlatform is working to standardize the TEE. Standardizing the TEE is crucial for mobile wallets, NFC payment implementations, premium content protection and bring your own device (BYOD) initiatives.

These following TEE specifications are currently available from the GlobalPlatform website:

- TEE Client API Specification v1.0 outlines the communication between applications running in a mobile OS and trusted applications residing in the TEE.
- TEE Systems Architecture v1.0 explains the hardware and software architectures behind the TEE.
- TEE Internal API Specification v1.0 specifies how to develop trusted applications.
- TEE Secure Element API Specification v1.0 specifies the syntax and semantics of the TEE Secure Element API. It is suitable for software developers implementing trusted applications running inside the TEE which need to expose an externally visible interface to client applications.
- Trusted User Interface API Specification v1.0 specifies how a trusted UI should facilitate information that will be securely configured by the end user and securely controlled by the TEE.
- TEE TA Debug Specification v1.0 specifies the GlobalPlatform TEE debug interfaces and protocols.

Joint venture formed by ARM, Gemalto and Giesecke & Devrient (G&D), Trustonic, was the first to qualify a GlobalPlatform-compliant TEE product in 2013.

## Security

The GlobalPlatform TEE Protection Profile specifies the typical threats the hardware and software of the TEE needs to withstand. It also details the security objectives that are to be met in order to counter these threats and the security functional requirements that a TEE will have to comply with. A security assurance level of EAL2+ has been selected; the focus is on vulnerabilities that are subject to widespread, software-based exploitation.

The Common Criteria portal has officially listed the GlobalPlatform TEE Protection Profile on its website, under the Trusted Computing category. This important milestone means that industries using TEE technology to deliver services such as premium content and mobile wallets, or enterprises and governments establishing secure mobility solutions, can now formally request that TEE products are certified against this security framework.

GlobalPlatform is committed to ensuring a standardized level of security for embedded applications on secure chip technology. It has developed an open and thoroughly evaluated trusted execution environment (TEE) ecosystem with accredited laboratories and evaluated products. This certification scheme created to certify a TEE product in 3 months has been launched officially in June 2015.

## Mobile Phone Tracking

Mobile phone tracking is the ascertaining of the position or location of a mobile phone, whether stationary or moving. Localization may occur either via multilateration of radio signals between (several) cell towers of the network and the phone, or simply via GPS. To locate a mobile phone using multilateration of radio signals, it must emit at least the roaming signal to contact the next nearby antenna tower, but the process does not require an active call. The Global System for Mobile Communications (GSM) is based on the phone's signal strength to nearby antenna masts.

Mobile positioning may include location-based services that disclose the actual coordinates of a mobile phone, which is a technology used by telecommunication companies to approximate the location of a mobile phone, and thereby also its user.

## Technology

The technology of locating is based on measuring power levels and antenna patterns and uses the concept that a powered mobile phone always communicates wirelessly with one of the closest base stations, so knowledge of the location of the base station implies the cell phone is nearby.

Advanced systems determine the sector in which the mobile phone is located and roughly estimate also the distance to the base station. Further approximation can be done by interpolating signals between adjacent antenna towers. Qualified services may achieve a precision of down to 50 meters in urban areas where mobile traffic and density of antenna towers (base stations) is sufficiently high. Rural and desolate areas may see miles between base stations and therefore determine locations less precisely.

GSM localization uses multilateration to determine the location of GSM mobile phones, or dedicated trackers, usually with the intent to locate the user.

The location of a mobile phone can be determined in a number of ways:

### Network-based

The location of a mobile phone can be determined using the service provider's network infrastructure. The advantage of network-based techniques, from a service provider's point of view, is that they can be implemented non-intrusively without affecting handsets. Network-based techniques were developed many years prior to the widespread availability of GPS on handsets.

The accuracy of network-based techniques varies, with cell identification as the least accurate and triangulation as moderately accurate, and newer "advanced forward link trilateration" timing methods as the most accurate. The accuracy of network-based techniques is both dependent on the concentration of cell base stations, with urban environments achieving the highest possible accuracy because of the higher number of cell towers, and the implementation of the most current timing methods.

One of the key challenges of network-based techniques is the requirement to work closely with the service provider, as it entails the installation of hardware and software within the operator's infrastructure. Frequently the compulsion associated with a legislative framework, such as Enhanced 9-1-1, is required before a service provider will deploy a solution.

## Handset-based

The location of a mobile phone can be determined using client software installed on the handset. This technique determines the location of the handset by putting its location by cell identification, signal strengths of the home and neighboring cells, which is continuously sent to the carrier. In addition, if the handset is also equipped with GPS then significantly more precise location information can be then sent from the handset to the carrier.

Another approach is to use a fingerprinting-based technique, where the "signature" of the home and neighboring cells signal strengths at different points in the area of interest is recorded by war-driving and matched in real-time to determine the handset location. This is usually performed independent from the carrier.

The key disadvantage of handset-based techniques, from service provider's point of view, is the necessity of installing software on the handset. It requires the active cooperation of the mobile subscriber as well as software that must be able to handle the different operating systems of the handsets. Typically, smartphones, such as one based on Symbian, Windows Mobile, Windows Phone, BlackBerry OS, iOS, or Android, would be able to run such software, e.g. Google Maps.

One proposed work-around is the installation of embedded hardware or software on the handset by the manufacturers, e.g., Enhanced Observed Time Difference (E-OTD). This avenue has not made significant headway, due to the difficulty of convincing different manufacturers to cooperate on a common mechanism and to address the cost issue. Another difficulty would be to address the issue of foreign handsets that are roaming in the network.

## SIM-based

Using the subscriber identity module (SIM) in GSM and Universal Mobile Telecommunications System (UMTS) handsets, it is possible to obtain raw radio measurements from the handset. Available measurements include the serving Cell ID, round-trip time, and signal strength. The type of information obtained via the SIM can differ from that which is available from the handset. For example, it may not be possible to obtain any raw measurements from the handset directly, yet still obtain measurements via the SIM.

### Wi-Fi

Crowdsourced Wi-Fi data can also be used to identify a handset's location. Poor performance of the GPS-based methods in indoor environment and increasing popularity of Wi-Fi have encouraged companies to design new and feasible methods to carry out Wi-Fi-based indoor positioning. Most smartphones combine Global Navigation Satellite Systems (GNSS), such as GPS and GLONASS, with Wi-Fi positioning systems.

### Hybrid

Hybrid positioning systems use a combination of network-based and handset-based technologies for location determination. One example would be some modes of Assisted GPS, which can both use GPS and network information to compute the location. Both types of data are thus used by the telephone to make the location more accurate (i.e., A-GPS). Alternatively tracking with both systems can also occur by having the phone attain its GPS-location directly from the satellites, and then having the information sent via the network to the person that is trying to locate the telephone. Such systems include Google Maps, as well as, LTE's OTDOA and E-CellID.

There are also hybrid positioning systems which combine several different location approaches to position mobile devices by Wi-Fi, WiMAX, GSM, LTE, IP addresses, and network environment data.

### Operational Purpose

In order to route calls to a phone, the cell towers listen for a signal sent from the phone and negotiate which tower is best able to communicate with the phone. As the phone changes location, the antenna towers monitor the signal, and the phone is "roamed" to an adjacent tower as appropriate. By comparing the relative signal strength from multiple antenna towers, a general location of a phone can be roughly determined. Other means make use of the antenna pattern, which supports angular determination and phase discrimination.

Newer phones may also allow the tracking of the phone even when turned on and not active in a telephone call. This results from the roaming procedures that perform hand-over of the phone from one base station to another.

### Bearer Interest

A phone's location can be uploaded to a common website where one's friends and family can view one's last reported position. Newer phones may have built-in GPS receivers which could be used in a similar fashion, but with much higher accuracy. This is controversial, because data on a common website means people who are not "friends and family" may be able to view the information.

### Privacy

Locating or positioning touches upon delicate privacy issues, since it enables someone to check where a person is without the person's consent. Strict ethics and security measures are strongly recommended for services that employ positioning. In 2012 Malte Spitz held a TED talk on the issue of mobile phone privacy in which he showcased his own stored data that he received from Deutsche Telekom after suing the company. He described the data, which consists of 35,830 lines of data collected during the span of Germany's data retention at the time, saying, "This is six months of my life [...] You can see where I am, when I sleep at night, what I'm doing." He partnered up with ZEIT Online and made his information publicly available in an interactive map which allows users to watch his entire movements during that time in fast-forward. Spitz concluded that technology consumers are the key to challenging privacy norms in today's society who "have to fight for self determination in the digital age."

## China

China has proposed using this technology to track commuting patterns of Beijing city residents. Aggregate presence of mobile phone users could be tracked in a privacy-preserving fashion.

## Europe

In Europe most countries have a constitutional guarantee on the secrecy of correspondence, and location data obtained from mobile phone networks is usually given the same protection as the communication itself.

## United States

In the United States, there is no explicit constitutional guarantee on the privacy of telecommunications, so use of location data is limited by law. Law enforcement can obtain permission to position phones in emergencies where people, including criminals, are missing. In some instances, law enforcement may even access a mobile phone's internal microphone to eavesdrop on conversations while the phone is switched off.

A secret interpretation of The Patriot Act, confirmed to exist, has been linked to secret widespread location tracking.

Since 2005 the Electronic Frontier Foundation has been following some U.S. cases, including *USA v. Pen Register*, regarding government tracking of individuals. In *In re Application of the United States for Historical Cell Site Data*, 724 F.3d 600 (5th Cir. 2013), the United States Court of Appeals for the Fifth Circuit held that the government does not need a warrant to compel cell phone providers to disclose historical cell site information. However, in *United States v. Davis (2014)*, the United States Court of Appeals for the Eleventh Circuit ruled in a criminal case that obtaining cell phone location data "without a warrant is a Fourth Amendment violation."

In 2014, it was revealed that in order to find fugitives, the United States Marshals Service has been flying small aircraft with equipment that identifies all cell phones in the area.

### Commercial Privacy of Location Information in the United States

The U.S. does limit commercial use of location information under the (US) Telecommunications Act, at 47 CFR §222. The Telecommunications Act, at 47 CFR §222(f), requires consent from the subscriber, and prohibits telecommunication common carriers from accessing location information for purposes other than system operation without consent of the customer. Businesses such as Locaid, which provide a tracking service based on subscriber information, require mobile users' consent prior to tracking.

## References


- Hantula, Richard (2010). How Do Cell Phones Work?. Google Books. 132 West 31st Street, New York NY 10001, US: Infobase Publishing. p. 27. ISBN 978-1-43812-805-4. Retrieved 18, January 2020
- Paganini, Pierluigi (March 7, 2014). "Dendroid – A new Android RAT available on the underground". securityaffairs.co. Retrieved 08, March 2020
- Kirat, Dhilung; Vigna, Giovanni; Kruegel, Christopher (2014). Barecloud: bare-metal analysis-based evasive malware detection. ACM. pp. 287–301. ISBN 978-1-931971-15-7
- Drake, Joshua; Lanier, Zach; Mulliner, Collin; Fora, Pau; Ridley, Stephen; Wicherski, Georg (March 2014). Android Hacker's Handbook. Wiley. p. 576. ISBN 978-1-118-60864-7
- Christopher Elisan (5 September 2012). Malware, Rootkits & Botnets A Beginner's Guide. McGraw Hill Professional. pp. 10–. ISBN 978-0-07-179205-9
- Dunham, Ken; Abu Nimeh, Saeed; Becher, Michael (2008). Mobile Malware Attack and Defense. Syngress Media. ISBN 978-1-59749-298-0

# Permissions

All chapters in this book are published with permission under the Creative Commons Attribution Share Alike License or equivalent. Every chapter published in this book has been scrutinized by our experts. Their significance has been extensively debated. The topics covered herein carry significant information for a comprehensive understanding. They may even be implemented as practical applications or may be referred to as a beginning point for further studies.

We would like to thank the editorial team for lending their expertise to make the book truly unique. They have played a crucial role in the development of this book. Without their invaluable contributions this book wouldn't have been possible. They have made vital efforts to compile up to date information on the varied aspects of this subject to make this book a valuable addition to the collection of many professionals and students.

This book was conceptualized with the vision of imparting up-to-date and integrated information in this field. To ensure the same, a matchless editorial board was set up. Every individual on the board went through rigorous rounds of assessment to prove their worth. After which they invested a large part of their time researching and compiling the most relevant data for our readers.

The editorial board has been involved in producing this book since its inception. They have spent rigorous hours researching and exploring the diverse topics which have resulted in the successful publishing of this book. They have passed on their knowledge of decades through this book. To expedite this challenging task, the publisher supported the team at every step. A small team of assistant editors was also appointed to further simplify the editing procedure and attain best results for the readers.

Apart from the editorial board, the designing team has also invested a significant amount of their time in understanding the subject and creating the most relevant covers. They scrutinized every image to scout for the most suitable representation of the subject and create an appropriate cover for the book.

The publishing team has been an ardent support to the editorial, designing and production team. Their endless efforts to recruit the best for this project, has resulted in the accomplishment of this book. They are a veteran in the field of academics and their pool of knowledge is as vast as their experience in printing. Their expertise and guidance has proved useful at every step. Their uncompromising quality standards have made this book an exceptional effort. Their encouragement from time to time has been an inspiration for everyone.

The publisher and the editorial board hope that this book will prove to be a valuable piece of knowledge for students, practitioners and scholars across the globe.

# Index

www.ingramcontent.com/pod-product-compliance
Lightning Source LLC
LaVergne TN
LVHW080717210926
848485LV00003B/477